Lenoir, NC
1994

Cable Television
Proof-of-Performance

A Practical Guide to Cable TV
Compliance Measurements
Using a Spectrum Analyzer

Cable Television Proof-of-Performance

A Practical Guide to Cable TV Compliance Measurements Using a Spectrum Analyzer

Jeffrey L. Thomas

For book and bookstore information

http://www.prenhall.com
gopher to gopher.prenhall.com

Prentice Hall PTR
Upper Saddle River, New Jersey 07458

Library of Congress Cataloging-in-Publication Data

Thomas, Jeffrey L.
    Cable television proof-of-performance : a practical guide to cable
TV compliance measurements using a spectrum analyzer / by Jeffrey L.
Thomas.
        p. cm. -- (Hewlett-Packard professional books)
    ISBN 0-13-306382-8
    1. Cable television--Testing. 2. Spectrum analyzers. I. Title.
II. Series.
TK6675.T56 1995
621.388'57'0287--dc20
                                        95-3896
                                          CIP

Editorial/Production Supervision: Ann Sullivan
Cover Design: Miguel Ortiz
Buyer: Karen Gettman
Editorial Assistant: Barbara Alfieri

Published by Prentice Hall PTR
Prentice-Hall Inc.
Upper Saddle River, New Jersey 07458

Definitions for selected terms in this book's glossary are derived from the *Jones Dictionary of Cable Television Terminology,* 3rd. (Glenn R. Jones, 1988, Jones 21st Century, Inc.), and are used with permission of the publisher

The publisher offers discounts on this book when ordered in bulk quantities.
For more information, contact:

Corporate Sales Department
Prentice Hall PTR
1 Lake Street
Upper Saddle River, New Jersey 07458

Phone: 800-382-3419, 201-236-7148
Fax: 201-236-7141
email: dan_rush@prenhall.

Printed in the United States of America

10 9 8 7

ISBN   0-13-306382-8

Prentice-Hall International (UK) Limited, *London*
Prentice-Hall of Australia Pty. Limited, *Sydney*
Prentice-Hall Canada Inc., *Toronto*
Prentice-Hall Hispanoamericana, S.A., *Mexico*
Prentice-Hall of India Private Limited, *New Delhi*
Prentice-Hall of Japan, Inc., *Tokyo*
Prentice-Hall Asia Pte. Ltd., *Singapore*
Editora Prentice-Hall do Brasil, Ltda., *Rio de Janeiro*

To Celeste, for her loving confidence.

Table of Contents

Preface

Your job as a cable television technician or engineer continues to change as your cable system evolves. The last ten years have seen a doubling in the number of subscribers and a ten-fold increase in the services offered. These changes demand increased technical performance from your cable system and an increase in the complexity and frequency of system testing. This book will help you keep up with test and technology growth by teaching you the basics of broadband RF cable compliance and maintenance measurements using a general-purpose spectrum analyzer.

Common industry test procedures, including those developed in Hewlett-Packard's Santa Rosa cable television research and design lab, have been used. The spectrum analyzer, with its proven utility to the cable industry, provides quality measurements while giving you a graphic vision of the measurement results. It is the only single instrument capable of full compliance testing. I have used Hewlett-Packard spectrum analyzers, but have kept product-specific controls and features to a minimum in order to widen the book's versatility and lengthen its useful life.

The chapters are organized by measurement complexity, progressing from carrier amplitude and frequency, to the more complicated noise- and distortion-related testing. Interference, leakage, audio, and video measurements round out the final chapters. Appendices provide background references, including a glossary of common cable TV and spectrum analyzer terms, and a tutorial on differential gain, differential phase, and chrominance-to-luminance delay inequality.

In selling services in today's marketplace, the quality of service and efficiency of testing maximizes your company's return-on-investment. Therefore, where possible, emphasis is placed on making tests that do not interfere with the delivery of services to your subscribers. Making tests efficiently and without disrupting product delivery will become more important as competing entertainment comes knocking at your customers' information door.

I will feel successful about the work put into this book if, after applying the book's concepts, you find that your cable maintenance and proof testing jobs make more sense, are more efficient, and are a bit more fun.

Jeff Thomas
Internet jeff_thomas@hp.com
CompuServe 74211,2014

Acknowledgments

I want to thank Francis Edgington of the Hewlett-Packard Company at Santa Rosa and Randy Goehler of Cox Cable Television of San Diego, Inc., both in California, for their willing contributions to and review of this book. Other people at the Hewlett-Packard who guided its direction and content were Rex Bullinger, John Cecil, Bill Koerner, and Bruce McPherran. Thanks goes to Larry Stratford of the HP Santa Rosa division for sparking the concept and heading the sponsorship of this book; Larry was responsible for getting me to write my first technical manual fifteen years ago. Thanks also to Duane Hartley and Jim Simpson, HP division management, for their support; Pat Pekary, of HP Press, and Karen Gettman, of Prentice Hall, for their enthusiastic guidance; and Lynn Denley-Bussard for her tireless attention to detail and inventive desk-top publishing skills.

1

Introduction

Overview

This book will teach you how to make cable television system compliance measurements with a spectrum analyzer. Its text, graphics, and examples provide a background in measurement technology, along with signal and test theory. Don't let the word theory intimidate you. The emphasis is on the practical principles of television signals, measurement concepts, and system carriers, not their detailed mathematical analysis.

*In this chapter
you'll learn about:*

- Working examples
- Tips & cautions
- Prerequisites
- Spectrum analyzer basics

What You Will Learn

You will learn to make the proof-of-performance measurements with a spectrum analyzer, whether in the relative comfort of the head end, or in the field where time, temperature, and weather make efficient and accurate measurements important.

The measurements covered are for NTSC-specified signals with compliance to the U.S. Federal Communications Commission (FCC) Rules and Regulations for cable television systems. Quotes from these regulations are in their respective chapters. Many of the techniques and guidelines apply to systems with PAL television standard signals since the modulation is similar, although channel and carrier spacing is different from NTSC, and wider modulation bandwidths require different measurement settings on the spectrum analyzer.

Working Examples

The examples in this book show step-by-step measurement procedures for compliance tests, including the amplitude and frequency of carriers, the hunt for intermodulation and cross modulation distortion products, the diagnosis of hum disturbances, and practical and system flatness tests. Each example is supported with material to help you understand the objectives, not just the law, of the pertinent compliance test, and how to get the best accuracy from the spectrum analyzer based upon measurement procedure and the analyzer's own data sheet specifications.

Learning is effective when reinforced with hands-on experience. This text can be used as the basis for a working course. Its examples serve as laboratory experiments when guided by a skilled cable television measurement practitioner. The examples have been written without using the key stroke sequences of any one specific spectrum analyzer model, although specifications, and features of Hewlett-Packard spectrum analyzers are prominent.

Look for Tips and Cautions

Tips and cautions are highlighted to help summarize the measurement. A tip puts the main point of the message in a few, large typeface words to help you stay on track. Cautions provide guidance to help avert common measurement or interpretations mistakes. They look like this:

Tips provide a summary of the local text.

Provides a warning about common measurement errors.

Prerequisites

This course is intended for cable television technicians and engineers who have some experience with system and head end measurements. The information that appears to be reviews can be skipped, but will be valuable to those who have less measurement experience.

Knowledge of the spectrum analyzer and how it works is not a necessary prerequisite to learn from this book. The cable television test procedures will teach you all you need to know about spectrum analyzer operation, measurement circuits, specifications, and features. An appendix is provided to those of you who must see the innards of a spectrum analyzer. References to the analyzer's block diagram and components shown in this appendix will be cited from time to time in the text.

What You Can Do After StudyingThis Book

After studying and working through the examples of this book, you will be able to make a number of proof-of-performance measurements faster and with confidence in the accuracy of your measurement. Here is a sample of the tests covered: visual carrier frequency and amplitude, aural signal frequency and amplitude, FM deviation, system noise level, carrier-to-noise ratio, depth of modulation, cross modulation, CTB/CSO, hum, in-channel frequency response, and system frequency response.

What You Need to Know About a Spectrum Analyzer

After saying that you don't need to know much about a spectrum analyzer, here is a primer for those of you who have never laid hands on one. The rest of you can skip to the next section!

TIP   **The spectrum analyzer is just an ever-tuning radio receiver with a screen for viewing signals.**

The spectrum analyzer is simply a receiver, like an ever-tuning radio, which displays signal amplitude over frequency. Think of it as a television receiver when the viewer has a finger pressing on the channel Up button of the remote control. The spectrum analyzer, unlike the television receiver, doesn't skip from frequency to frequency; it tunes continuously across the selected frequency span, and then starts over. It does not display a result of the signal at its input, such as a TV picture, but the signal's amplitude and frequency components. It displays any signal in its input range, as illustrated in Figure 1. The amplitudes and frequencies of the harmonics of the left-most signal are displayed.

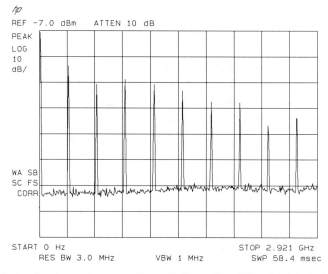

Figure 1. A spectrum analyzer look at its calibration signal, at 300 MHz, and all its harmonics.

What a Spectrum Analyzer Does

The spectrum analyzer is calibrated specifically for the accurate display of continuous wave (CW) signals, even though there are few such signals in the world of communications and television. CW signals do not contain information by themselves, but can carry a great deal of information in their modulation. This modulation can be broken down into still further sidebands that appear as CW signals. And the analyzer can display these!

This is an oversimplification of the use of a spectrum analyzer; however, this basic understanding of the spectrum analyzer underlies all the procedures within this book.

TIP	**Modulation of a CW signal can be viewed easily with the spectrum analyzer as sidebands.**

How do you control what the analyzer does? Simple. The spectrum analyzer has inputs and outputs, a cathode ray tube (CRT) display, and keys to access its functions. The functions are grouped for convenience, the primary ones controlling the CRT display of frequency and amplitude ranges. The preset key sets the analyzer to a known state. It is a safe and convenient reset button when you get confused about a measurement, when the analyzer decides to misbehave, or when you need a known starting point. Most spectrum analyzers will preset when powered on. Figure 2 shows a representative spectrum analyzer front panel.

The modern spectrum analyzer, as with most contemporary test equipment, has an internal computer or microprocessor which allows the instrument to do much more at the touch of a single control. As an operator, you control the analyzer hardware and circuits through its computer by issuing commands to its microprocessor by pressing or turning front panel controls. Figure 2 illustrates three types of controls called hardkeys, softkeys, and data entry controls. Hardkeys, often with a dedicated name on or near the key, enable the same function every time they are pressed. Softkeys, usually located adjacent to the CRT, may change definition depending upon the hardkey pressed. Softkey labels are shown on the CRT. Data controls are the number pad, unit keys, knob, and up and down keys. They are used to enter and change numeric settings by typing a number, or by stepping up or down through a predefined sequence.

Figure 3 shows a typical spectrum analyzer CRT display. The trace is annotated with up-to-the-minute analyzer amplitude and frequency states, control settings, and measurement messages. Amplitude level is given for the reference level, the top graticule, and a scale such as decibel (dB) per division. Frequency is shown as a center frequency

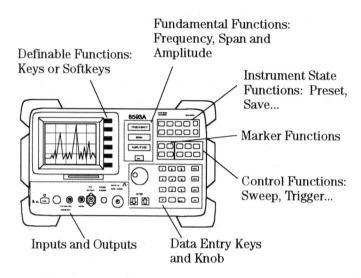

Figure 2. A typical spectrum analyzer front panel.

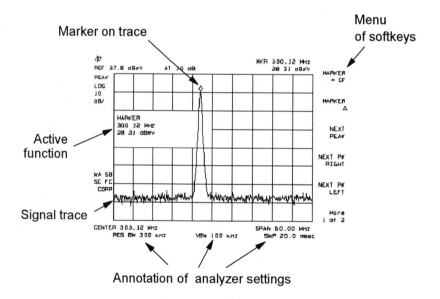

Figure 3. The display communicates all readout information. Other outputs include rear panel RF connectors, the loudspeaker, printer and plotter ports, and a computer port.

and as either a total span or start and stop frequencies. Note that frequency is given as full span rather than as a per division number as in older spectrum analyzers and most current oscilloscopes.

Promote Your Skill, Understanding, and Interest

The aim of this book is to make you more knowledgeable in your job as an operator and technical professional in the maintenance of your cable television system. The background and skills taught will help you prevent subscriber dissatisfaction, satisfy the letter and intent of the compliance regulations, and provide you with more insight into trouble-shooting performance problems. It is hoped that the material here will assist in your professional growth. This growth will be necessary to meet future measurement challenges as cable systems add more services, move toward digital signal processing, and expand to a wide two-way data highway. Along the way, perhaps, the tools and understanding this book can provide will reduce the amount of stress in your job, helping you to have more time and energy for a bit of technical exploration...in other words, some fun!

Selected Bibliography

- *Code of Federal Regulations, Title 47, Telecommunications, Part 76, Cable Television Service.* Federal Commission Rules and Regulations, 1990.
- Peterson, Blake. *Spectrum Analysis Basics.* Hewlett-Packard Company, Application Note AN 150, Literature No. 5952-0292, Santa Rosa, CA, 1989.

Through the Spectrum Analyzer Looking Glass

2

Overview

This chapter introduces you to the spectrum analyzer by showing you how it views parts of the system spectrum through its "looking-glass," the CRT display, with examples to demonstrate measurement ease and versatility. The attributes of a analyzer required for cable measurements are outlined to help you select one appropriate for your testing needs. Finally, compliance test methodology is reviewed.

In this chapter you'll learn about:

- System overview
- Technical standards
- Methodology
- Regulations
- Spectrum Analyzer features

System Overview

System Overview Provides Comfort, Not Test Results

To get familiar with the spectrum analyzer, let's look at an overview of an entire system's frequency range.

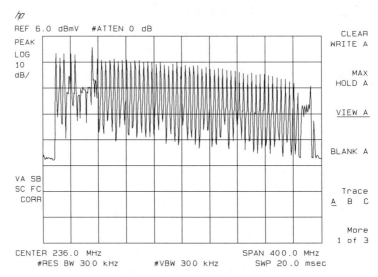

Figure 4. A standard frequency configuration with 54 channels and commercial FM shown on the spectrum analyzer.

A view of all the channels at once provides a quick look at your system's signature, that is, its total spectrum response. Don't expect to get much specific measurement information from this view; it's just for observing the general health of the system, that is, detecting major performance variations or disruptions. The example below will help you get familiarwith the spectrum analyzer controls, its ease of use, and measurement speed. The first example enables you to see the whole system.

Example 1. Displaying the Full System

Bring the spectrum analyzer to a known state by pressing its Preset key, or by cycling the line power off and on again. Usually this start-up instrument condition selects the

---

The Measurement in Brief

1. Turn the analyzer on and connect the cable drop to the RF input.
2. Press the Preset to start from a known setup.
3. Set the frequency start and stop for the range of your system, 50 MHz to 400 MHz, for example.
4. Use the reference level control to bring the signal amplitudes higher on the display.
5. Use the marker controls to point and read the amplitude and frequency of different signals.

---

analyzer's widest frequency range, so the signals are seen as a clump of amplitude responses towards the left, or lower frequency side, as shown in Figure 5. The frequencies are so close together that they appear as one large signal.

One additional signal is at the far left boundary of the display. This is not a signal in the cable system. It is an artifact produced by the spectrum analyzer itself, known as the LO feedthrough. It marks the 0 Hz frequency point. More on the origin of this response can be found in the spectrum analyzer basics reference.

Now change the frequency span of the spectrum analyzer so that the display closes in on the upper and lower system frequency edges. In this case the system signals are within 30 to 450 MHz.

Spectrum analyzers have a frequency menu that usually includes a start and stop frequency. If your analyzer has these, simply enter the two values, being careful to enter the smaller frequency as the start, and the higher frequency as the stop. If the analyzer does not have a start and stop frequency control, enter a center frequency midway between the desired start and stop, that is $(450-30)/2 + 30$ MHz $=$ 240 MHz and a span equal to the difference between the start and stop frequencies, $450 - 30 =$ 420 MHz.

Now bring the highest signal responses near to the top graticule with the amplitude or reference level control. The reference level now represents the highest signal levels. The display in Figure 4 results.

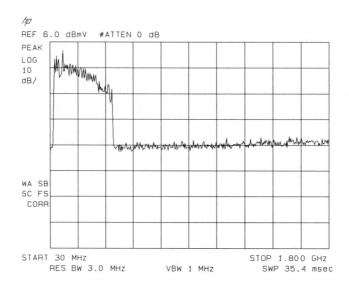

Figure 5. The spectrum of a cable television system viewed on a 1 MHz to 1.8 GHz spectrum analyzer CRT in full span.

As stated before, this measurement is not very accurate; that is, although the visual carriers can clearly be seen, the analyzer is set to too wide a span for compliance-level accuracy.

TIP   **Use the analyzer to look at the entire system frequency response at once.**

Taking the span narrower to just a few channels' width, as in Figure 6, demonstrates some important aspects of the spectrum analyzer. These are the ability:

◆ To show large and small signals at the same time for comparison.
◆ To make absolute signal strength and frequency measurements directly on individual signals.
◆ To identify and measure modulation on a signal.

All these are critical to cable television compliance measurements. Examples 2 and 3 illustrate the ease with which detailed information about the signals in the system can be obtained with the spectrum analyzer simply by adjusting three of its controls: center frequency, frequency span, and reference level.

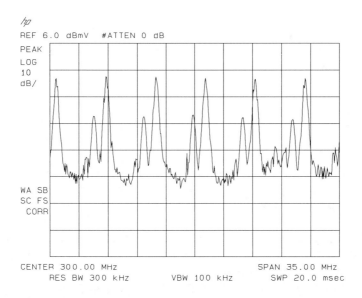

Figure 6. Narrowing the analyzer's frequency span to 35 MHz, centered among the TV channels shows how much more information becomes available as the frequency span is reduced.

Example 2. Looking at a Few Channels

Starting from the last example, select a center frequency of 138 MHz and a span of 35 MHz. Note that the signals become even more distinct from one another as the span is narrowed. More information about the spectrum is available; the channel aural carriers are now visible. The spectrum analyzer automatically adjusted its signal selectivity in proportion to the frequency span to present a view of individual signals.

Individual carriers can be seen, but the signals don't look stable. Their apparent instability in amplitude and frequency is caused by the modulation. Now we are in close enough to start making measurements. Since each signal can be seen distinct from its neighbor, signal levels can be compared, a major strength of the spectrum analyzer. About 90 percent of cable measurements are comparisons of one signal to another.

TIP	**Most measurements are comparisons between signal amplitudes and frequencies.**

Expanding one aural carrier to the analyzer's full display illustrates the third analyzer strength—its ability to identify modulation.

Example 3. Observing FM Aural Carrier

From the spectrum analyzer state in the last example, place one of the aural carriers in the center of the analyzer's display and reduce the frequency span with the Step Down key. The signal appears to drift to one side or the other as the analyzer expands the frequency spectrum about the center of the display. You may have to readjust the carrier to the center with the center frequency control.

Step down the full span to 100 kHz, and bring the level of the signal close to the top graticule with the reference level control. When there is no sound on the channel, the carrier appears as a single spike, a CW signal. As the carrier is moving back and forth at a rate and distance dictated by the sound modulation, it is displayed as a splatter of amplitudes on each side of the carrier. During quiet times, the carrier can be seen at rest as a CW signal.

The spectrum analyzer is adept at identifying signal modulation. Further analysis and measurement of the FM carrier are discussed in the chapter on FM deviation.

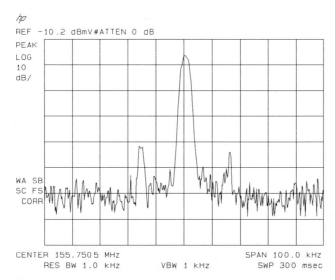

Figure 7a. The display of a single channel's aural carrier shows the carrier with little modulation.

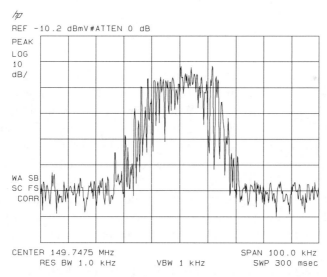

Figure 7b. This illustration shows the carrier with rock and roll music.

Technical Standards for Compliance Testing

Regulations are getting tougher because entertainment services must offer high quality to compete in a supply-driven market, and because of the increasing complexity of systems and the signals they carry. The FCC has had cable television rules in place for as long as most of us can remember. The 1980s brought little legislation toward increased performance testing, but the 1990s are proving different. Society's attention away from defense and aerospace, and the focus on the information and entertainment avenues of distribution, stimulated by competition, will keep regulations tightening through the turn of the century.

Changes recently have required new levels of testing and data gathering. The technical standards are set out in Part 76 of Title 47, the telecommunications section of the Federal Rules and Regulations. Part 76 technical standards were first adopted in 1972. In 1985, the FCC made most of the Part 76 technical standards "guidelines" rather than requirements. In February 1992, the FCC reinstated and updated the Part 76 technical standards requiring cable television systems to measure and record their signals at widely separated test points. For the first 3 years, only radiofrequency (RF) measurements are required. After that, demodulated baseband video measurements must be made on a 3-year-cycle basis as well.

Testing Methodology

Testing methodology means the what, when, how, where, and by whom of the tests. All NTSC video-based cable channels are subject to testing. RF tests, that is 76.605 (a) (1-10), are subject to testing twice a year. While compliance is required for all cable systems, those with fewer than 1,000 subscribers are exempted from the testing requirements. For cable systems with 1,000 to 12,500 subscribers, measurements must be taken at a minimum of six widely separated points at least one-third of which must be representative of subscriber terminals most distant from the system input. For systems over 12,500 subscribers, one test point needs to be added for every additional 12,500 subscribers. Records must be maintained for a period of 5 years, including instrumentation and procedures used and the qualifications of the person performing the testing.

Fiber hubs need to be included in the test points requirement—at least one test point for each hub. A fiber hub is considered a "remote" point. No specific testing on the fiber signal itself is called for.

Test points are located wherever there is a mini-head end, such as a fiber optic hub or AML receiver.

The regulations call for testing at the subscriber terminal. This means that all the testing should be done at the output of the set-top converter box in the customer's home, not very practical, and the FCC recognizes this. But the letter of the regulations have not changed, so an unwritten compromise has evolved. Converter boxes are type-tested, and the data is submitted along with the compliance test results taken at the end of 100 feet of cable from the subscriber's tap.

Meeting Regulations Using a Spectrum Analyzer

The spectrum analyzer is the recommended test instrument for most proof-of-performance measurements. Its versatility and performance continue to improve since its first popular application to cable measurements in the early 1970s. As cable measurements become more demanding, the spectrum analyzer has increased in performance, with competitive pressures keeping prices within the cable system budget, and providing a better measurement value. That's not to say that a spectrum analyzer is cheap! But its value is recognized and accepted by cable system operators around the world.

The spectrum analyzer can make compliance measurements for all the current tests. Some tests require special added options.

Selecting a Spectrum Analyzer

Presumably you purchased this book because you already have a spectrum analyzer and wish to squeeze the most performance out of it for your testing purposes. Here are the minimum performance requirements for testing your cable television system. Be advised that not all proof-of-performance measurements depend on raw data sheet specifications. Some tests require special features, such as very fast sweeps in the time domain.

A spectrum analyzer data sheet puts performance in terms that may not have been written with the cable industry in mind. Don't worry about understanding all of these now; the following measurement chapters will explain the specifications in relationship to the tests and the accuracy of the results. Also see the glossary at the end of the book, which summarizes these spectrum analyzer terms in relation to the cable industry.

- Frequency range: 10 to 1000 MHz
- Frequency spans: zero, and 100 kHz to 1000 MHz
- Frequency accuracy: ± 200 Hz
- Relative amplitude accuracy: ± 2.0 dB over entire frequency range, with enough detailed amplitude specifications to calculate the relative accuracy to ± 0.5 dB
- Maximum input level: 1 watt damage level, AC-coupled
- Sensitivity, the smallest signal that can be measured: −60 dBmV
- Noise floor, related to sensitivity: −60 dBmV in narrowest resolution bandwidth
- Internal distortion products: ≤60 dBc with total input to the analyzer's mixer 10 dBmV
- Resolution bandwidths: 1 kHz to 3 MHz, video receiver with 4 MHz video bandwidth
- Video bandwidths: equal to the resolution bandwidths
- Input attenuator: 0 to 60 dB in 10 dB steps or smaller
- Input preamplifier: internal or external >20 dB gain, <7 dB noise figure
- 75-ohm input impedance

Additional features make the analyzer easier, faster, and more accurate to use. Here are the main ones:

- Frequency and amplitude markers on the trace
- Analog CRT display
- Sweeps in zero span: 10 μsec full span
- Television sync trigger
- Fast Fourier transform on analyzer's detected output
- FM demodulator
- Television picture and/or sound on the analyzer's CRT
- Negative peak detection

These lists may look formidable, but many spectrum analyzers on the market today offer these capabilities and superior performance at a reasonable price.

What's To Come

The measurements in this book are organized so as to teach the more fundamental aspects of the spectrum analyzer first. With each new chapter, additional measurement concepts are introduced along with the performance, techniques, and features necessary to make the measurements. Cable measurements with similar test techniques are clustered as much as possible, but if the order doesn't always seem to make sense, skip around.

Selected Bibliography

- Engelson, Morris. *Modern Spectrum Analyzer Measurements*. Portland: published by JMS, 1991.
- *HP 85721A Cable TV Measurements and System Monitor Personality.* Hewlett-Packard Company, User's Guide, Part No. 85721-90001, December 1993.
- Peterson, Blake. *Spectrum Analysis Basics*. Hewlett-Packard Company, Application Note AN 150, Literature No. 5952-0292, Santa Rosa, CA, 1989.

3

System and Signal Review

Overview

The last chapter introduced the spectrum analyzer as the primary test instrument for proof-of-performance measurements. This chapter reviews the basic signal concepts necessary to understand the detailed measurement procedures that start in Chapter 4. Much or all of this will be reviewed. Have patience, though, the perspective of the cable television system signals in the frequency domain may be quite different than the view you are used to with other test equipment.

Where Measurements Are Made

The quality of the signals distributed throughout a cable television system is best at the point of origin. In the cable system this point is the head end. Factors that degrade the signal's quality include the attenuation of the distribution cable; the noise added by trunk, distribution, and feeder amplifiers; the "new" signals added as distortion products from amplification and mixing; modulation from power supplies; and encroaching signals from sources outside.

TIP   **Performance gets no better than at the head end.**

For you, the cable operator, the head end is the gathering point of signals that come from a variety of sources, each with its own signal path and processing. Each of these signals must be checked for quality before being passed along to the subscribers. Often viewing and listening to the baseband signal on a monitor is sufficient, but with increased emphasis on video quality legislated for the remainder of the 1990s, video measurements such as depth of modulation, FM deviation, differential phase and gain should also be made.

Head End Measurements

If you had to select the bare minimum of measurements to be made at the head end, they may look something like this:

- ◆ Amplitude and frequency of all visual carriers
- ◆ Carrier-to-noise ratioDepth of modulation
- ◆ FM deviation
- ◆ Channel frequency response
- ◆ Differential phase, differential gain, and chroma-lumina delay inequality

Some of these are required for compliance. But all are critical to your confidence that the signals launched into the system are of high enough quality to satisfy your subscriber.

The video-quality measurements are required for 1995. Although these measurements cannot be made with a spectrum analyzer as defined for the measurements in this book, a brief tutorial and the receiver requirements necessary for these video measurements are in Appendix E, Color-Video Measurements.

Distribution System Requires More Testing

The tests that need to be run in the distribution system include most of the above tests, depending upon the location, complaints, legislation, and phase of the moon, plus these:

- ◆ Coherent disturbances
- ◆ Low-frequency disturbances
- ◆ Co-channel/ingress
- ◆ System flatness

Television in the Frequency Domain

Let's look in detail at how the television signal appears in the time and frequency domains. The following figures review the composition of the NTSC channel signal in the following familiar terms: the signal as it comes off the air, a vertical time frame, two horizontal time frames, and the resulting television raster.

The television broadcast signal, whether it be NTSC, PAL or SECAM, is the most complex analog signal used in commercial communications. It is comprised of amplitude, frequency, phase, and pulse modulation fitted into a 6 MHz channel with a single sideband transmission process called a vestigial sideband.

TIP   **Most compliance measurements are made in the frequency domain.**

In Figure 8 the amplitude modulation is shown as an envelope on the RF carrier, symmetrically about a zero-voltage amplitude. When a television receiver looks at this carrier, it picks off the envelope and discards the RF carrier.

Figure 9 expands a portion of this envelope to show the vertical scan as a frame sync pulse. The square sync pulse provides bursts of RF energy at the peak carrier amplitude. This repetitive burst synchronizes the vertical scan on the television receiver at a 59.94 Hz rate.

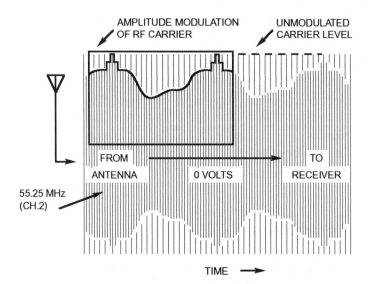

Figure 8. The visual carrier of a television signal as it appears demodulated over time. The maximum voltage swing represents the visual carrier level if it were unmodulated.

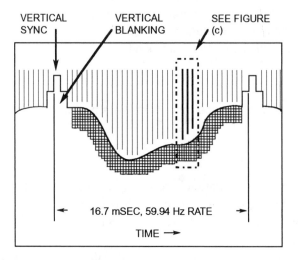

Figure 9. A close-up of one vertical frame in Figure 8. The vertical sync pulse returns the television beam to the top of the screen to start another field.

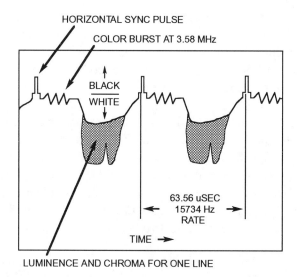

Figure 10. A close-up of two horizontal lines in Figure 9 shows the luminance, that is, black and white across each line of video.

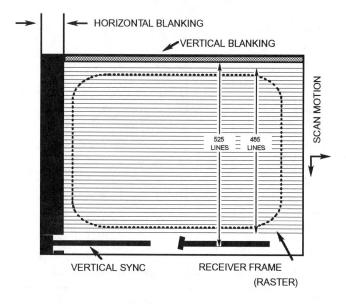

Figure 11. A television receiver raster, with the television picture in the dotted frame. The area outside the picture area shows synchronizing signals, additional programming information, and test signals.

Figure 10 is a 1:100 time-domain magnification of the waveform shown in Figure 9. It shows the horizontal sync pulses and luminance information on two of the 525 horizontal lines that make up a single frame. These sync pulses, which run at a 15,734.264 Hz rate, also have their tops at the peak RF carrier level. For color transmission this pulse includes a 3.579545 MHz burst on its trailing side.

Figure 11 shows the television picture that results from these modulations of the visual carrier. The area not viewed by a television receiver is also shown in the figure outside the dotted television mask. The horizontal black bar at the bottom of the screen is a group of vertical synchronizing pulses. The area at the top of the display contains supplemental program material, such as second audio program (SAP), subtitle text, tele-text, and video test signals used for on-line video performance tests that do not interfere with the regular programming. These test signals are called vertical interval test signals (VITS).

The above figures only show part of the television channel. The audio signal is missing in the time-domain pictures because it is a separate carrier transmitted along with the visual carrier as a sideband.

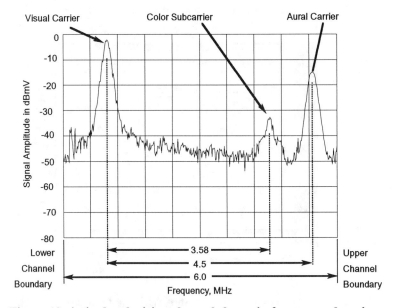

Figure 12. A single television channel shown in frequency domain. Its energy separates into sideband components: visual carrier, color subcarrier, and aural carrier.

After the baseband video signal is processed, the audio carrier is added by the modulator 4.5 MHz above the visual carrier and at less than one-eighth of its power for transmission down the cable.

Your Favorite TV Show in the Frequency Domain

Valuable information is available in a display of the signal's amplitude versus frequency. The vertical axis is scaled to the power of the signal, in units relative to a millivolt. The modulation shown as time domain signals in Figures 8 through 10 are displayed as sidebands in frequency in Figure 12. The largest signal is the visual carrier. The next largest is the audio carrier. Each is modulated separately by the transmitter and combined with the frequency spacing shown. Prior to the transmission the sidebands of the visual carrier, which are symmetrical on both sides to begin with, are filtered off the lower frequency side (to the left of the carrier) from 0.75 MHz on. This vestigial sideband technique is used to conserve the frequency spectrum, allowing all the modulation to be contained in a 6 MHz channel bandwidth. The television receiver uses the full upper sideband and a small portion of the filtered lower sideband to reconstruct the television video.

The chrominance, or color, carrier is 3.579545 MHz (3.58 MHz is detailed enough) above the carrier. This sideband contains the picture's color, or chrominance, information. Each horizontal sync pulse has its own 3.58 MHz burst to calibrate the television receiver's color circuitry for each horizontal line. The color information is phase modulated. The aural carrier is placed 4.50 MHz above the visual carrier, or 250 kHz from the upper edge of the channel. It is an FM signal with a 80 kHz bandwidth.

How Big is Big?

Signal power level is the key to system performance. The television receiver requires a certain level to reconstruct the television picture, and levels must be uniform from one end of the channel spectrum to the other, or television reception will be irregular. Visual carriers adjacent to one another must be close to the same level, or one will start imposing its modulation on the other. So the question is: How is a television channel level measured?

First, not all the channel sidebands need to be measured. The visual carrier contains so much of the channel energy that its strength is a consistent measure of a channel's power. This is why the compliance tests for absolute power levels use the visual carrier. Second, the visual carrier has complex modulation. How can a consistent power level be made? To answer that question, look at Figure 13. The modulation of the visual carrier is a special form of amplitude modulation where the modulation is down from the carrier level's maximum voltage swings. In other words, the peak of the video carrier is the voltage of the carrier as if it were not modulated.

Moving from left to right in Figure 13, (a) represents the visual carrier voltage as it would appear without modulation. The voltage swings occur at the rate of the carrier frequency. For example, the rate for channel 2 in a standard frequency allocation would be 55.25 MHz.

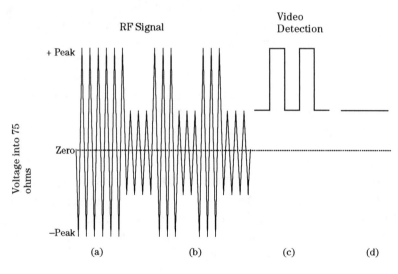

Figure 13. Amplitude modulation of carrier is down from the peak carrier voltage swing.

When video information is modulated onto the carrier between specified horizontal synch pulses, it leaves the amplitude of the pulse tops unchanged. This modulation scheme is represented by (b) in Figure 13.

The television receiver, waveform analyzer, and spectrum analyzer demodulate the visual carrier, represented by the square wave as shown in (c). The maximum values of this demodulated waveform represent the visual carrier peak power. Because pulse maximums are the only consistent signal levels related to the carrier's power and are easily extracted from the television signal, they are used to measure and compare television signal strengths.

TIP   **The carrier level is measured as the peaks of the sync tips.**

Convenient Measurement Units

The term dBmV is a familiar one. It is the standard unit adopted by the cable television industry to make absolute level measurements on all cable system. Here is why: One millivolt across 75 Ω is the energy of a strong signal at the cable or antenna input terminals of a television receiver, so it would be convenient to make power level and power differential measurements related to millivolts. Calculating power level and power differences in watts with voltage units is awkward at best. Power level requires the formula $V^2/75$, where V is the voltage across 75 Ω, to be computed each time. A power change means that the formula $(V_1^2 - V_2^2)/75$ has to be calculated. The decibel (dB) resolves these difficulties in handling system power figures by using a simple mathematical relationship between the powers to eliminate the need for very large or very small numbers when referring to relative or absolute powers. How does the millivolt get back into the picture? The dB will be defined, and then related to the millivolt as the familiar dBmV in the next paragraphs.

TIP   **All units with dB as part of their name represent a measure of power. dBmV is a unit that gives power referenced to a millivolt.**

The dB Always Represents Power Ratio

The number of dB's always represents a power ratio. Zero dB is a power ratio of one, meaning that the powers are equal. Ten dB is a power ratio of 10, meaning that one power is ten times the power of the other. Decibel is defined with the formula:

$$dB = 10 \log_{10}(P_T / P_R)$$
Equation 1

where P_T and P_R are the test power and the reference power, respectively.

The units of the powers are not important because they cancel each other. In other words, the ratio (P_T /P_R) has no units. If P_T is greater than P_R, then the dB value is a positive number, indicating a power gain. If P_T is less than P_R, then the dB value is a negative number, indicating a power loss.

dB ratios are easy to remember: 10 dB is a change of power by a factor of 10; and 3 dB is a change of 2; and 0 dB is no power change, that is, a factor of 1.

To find the power ratio when the number of dB is known, Equation 1 needs to be turned inside out, that is, solved for the power ratio as a function of the number of dB:

$$P_T /P_R = \log_{10}^{-1} (dB/10)$$
Equation 2

where P_T is the power value to be compared to the reference, P_R

Note that the value of dB carries its sign, + or −. That is critical in determining which signal is higher.

Table 1 below shows decibel values related to power ratio.

Table 1. Relationship between dB and power ratio.

dB	Power Ratio
−20	0.01
−10	0.10
−5	0.316
−3	0.50
−1	0.79
0	1.00
+1	1.26
+3	2.00
+5	3.16
+10	10.00
+20	100.00

Example 4. Calculation of dB from Power Ratio

The visual carrier is 13 dB above the aural carrier, that is +13 dB. What is the ratio of powers between the two signals?

This can be calculated with the formula or reconstructed from the table above. The visual carrier is the test power, P_T, that is, the power to be compared to the aural signal, the reference level, P_T. From the Equation 2, $P_T / P_R = \log_{10}^{-1} [+13/10] =$ 19.952 (20 is close enough). The visual carrier is 20 times more powerful than the aural carrier.

The same answer can be derived easily from the table of dB and ratios without the use of the formula. The +13 dB is comprised of +10 and +3. These ratios are 10 and 2, respectively. Since they are ratios, they are multiplied to give $10 \times 2 = 20$.

Simply remembering a few numbers from Table 1 will help you make conversions quickly. But more important, they give you a "feel" for the power ratios involved in your cable measurements. Amaze your friends!

If dB Is Power Ratio, What Is dBmV?

Decibel takes care of power ratios between signals. But measuring visual carrier levels at the subscriber terminal requires a unit for absolute power reading. Since the dB definition provides a convenient method of dealing with power differences, why not let a signal level be given in a dB referred to an absolute power? A dBW is power referred to a watt. A dBm is power relative to a milliwatt, or one one-thousandth of a watt. The number of dBW or dBm is an absolute power because it represents the difference from an absolute level.

Now, finally, let us relate all this back to the millivolt. The dBmV is the measure of power, a level relative to an absolute power. Since the impedance of the cable or television receiver is 75 Ω, the voltage represents the power using Ohm's law, Power = Voltage^2/75 Ω. The dB formula can be re-written for voltage ratio

$$dB = 10 \log_{10}(P_1 / P_2) = 20 \log_{10}(V_1 / V_2)$$
Equation 3

The difference between two values in dBW, dBm, or dBmV represents power ratios as given in Table 1.

Example 5. Calculating Power Differences

The visual carrier is −10 dBm and its neighboring carrier is +12 dBmV. What is the power ratio between the two signals?

A trick question. The two power levels are given in different units and cannot be subtracted directly. One of the values needs to be converted to units of the other before taking the difference.

From Appendix A, "Tables For Unit Conversion", 0 dBm is the equivalent power of +48.75 dBmV, so −10 dBm is +38.75 dBmV. The difference between the two adjacent carriers is 38.75 − 12 = 26.75 dB. The −10 dBm is the higher signal level.

Example 6. Use of the Spectrum Analyzer to Make Unit Conversions

The spectrum analyzer, used for applications where different measurement units are required, can convert units directly using its internal computer. To convert the −10 dBm value in the last example, set the analyzer's reference level to −10 dBm. Change the analyzer's units to dBmV, and the new reference level shows the conversion, +38.75 dBmV. This is the result of the addition of the conversion factor +48.75 dB to the −10 dBm power reading. Some analyzers round the value to a single decimal place because of readout accuracy considerations.

Matching the Spectrum Analyzer with the Cable Television System

Measurement accuracy depends upon a good impedance match between the spectrum analyzer and the cable system. The closer the test equipment impedance is to the system impedance the better the accuracy. Most modern spectrum analyzers offer an input impedance of 75 Ω, which eliminates match errors. But a majority of the spectrum analyzers sold to date for general use outside the cable industry have 50 Ω input impedance. A simple matching transformer or matching pad provides a suitable solution, as long as the insertion loss and frequency response of the matching device is taken in to account when making your measurements. Appendix A has more information.

> **CAUTION**   **Measurement errors can be caused by poor impedance matching between the cable and spectrum analyzer.**

Example 7. Calculation of Impedance Mismatch Error

What is the error and correction when a 50 Ω spectrum analyzer is connected to a 75 Ω cable?

From the impedance conversion table in Appendix A, the spectrum analyzer will read 1.76 dB lower than expected. The power lost is the result of an inefficient power transfer from a 75 Ω source to a 50 Ω load. In other words, the current from the cable (75 Ω) across the spectrum analyzer's smaller impedance (50 Ω) produces a lower voltage.

The computation in this example does not take into account the uncertainties caused by the gross mismatch of impedances on signals approaching 1000 MHz. At these high frequencies, reflections from the mismatch cause standing waves for measurement uncertainties of 5 to 15 dB.

Calibration of the Spectrum Analyzer

The final step in preparation for the measurements is to ensure that the spectrum analyzer is calibrated. The specifications for the analyzer's performance are based upon several factors:

- Operation for some minimum specified time to allow the internal temperatures to stabilize
- Calibration routines run and the data stored with a stabilized ambient temperature
- Use of an accurate and stable calibration signal often provided by the spectrum analyzer
- Application of the calibration factors during measurement

Your spectrum analyzer operation manual will guide you in the specific calibration procedures.

```
        BANDWIDTH                 AMPLITUDE

  6dB   BWAMP LC XTL    RFATN  SGAIN  LOG    LIN    NBW
200H   -0.01    0 134   -0.01   0dB -0.11
  9K    0.88  128 123    0.00  10dB  0.00   0.60   0.00
120K    0.72   46 255    0.05  20dB  0.34   0.50   0.00
                         0.04  30dB  0.44   0.78   0.00
  3dB   BWAMP LC XTL    -0.06  40dB  0.57   0.74   0.00
 10H    1.52  128 128   -0.02  50dB  0.68
 30H    1.09  108 108   -0.06  60dB
100H    0.89  128 128    0.00  70dB   MCDLY          90
300H    0.90  128 128
  1K    0.07    0 225    CA ATT ERR    BND GAIN ERR
  3K    0.01    0 169     1 -0.07      0 200 -0.11
 10K   -0.08    0 104     2 -0.04      1 200 -0.11
 30K   -0.18    0  33     4  0.15      2 200 -0.11
100K    0.53   59 255     8  0.11      3 200 -0.11
300K    0.05  111 255    16  0.00      4 200 -0.11
  1M   -0.16  170 255
  3M   -0.23  229 255
 ~5M   -0.31  244 255
                                              RL
```

Figure 14. An example of built-in calibration results.

TIP

For best accuracy, calibrate the spectrum analyzer after warm-up and whenever the ambient temperature changes more than 10° F.

Review

This chapter helped prepare the way for making measurements:

- Outline of the measurements to be made, and their relative importance to head end and distribution system performance

- A look at the television signal and how it appears in time and frequency domains
- The practical aspects of using a spectrum analyzer for measurements, including units, impedance, and calibration

Selected Bibliography

- *Cable Television System Measurement Handbook.* Hewlett-Packard Company, Literature No. 5952-9228, Santa Rosa, CA, January 1977.
- Benson, K. Blair, and Whitaker, Jerry. *Television Engineering Handbook.* Rev. ed., McGraw-Hill, Inc., 1992.

4

Measuring Carrier and Sideband Amplitude

Overview

Television channels are the mainstay of cable systems. Measurement of the performance of the cable system begins with the visual and aural carrier levels at the head end, because, without a uniform signal response through the system, few other measurements would be meaningful and customer complaints would come from all hubs and branches. Fortunately, the measurement of signal levels and frequencies are the simplest measurements made, and, therefore, the best measurements to begin our learning the use of the spectrum analyzer for cable system testing. First, the amplitude measurements, and then the frequency measurements are discussed.

In this chapter you'll learn about:

- Carrier amplitude measurement
- Relative and absolute
- Spectrum analyzer features
- Avoiding measurement mistakes

Absolute and Relative

Who ever said that everything is relative was almost correct. As mentioned before, a spectrum analyzer simultaneously measures absolute amplitude in dBmV and absolute frequency in Hz. The analyzer also measures the relative powers and frequencies between signals, in dB and Hz, respectively.

Figure 15 illustrates the concept as viewed on the spectrum analyzer display. Power levels of each signal, frequencies of each signal, and their differences are shown. The power levels and signal frequencies are absolute. The signal-level differences and frequency spacings are relative. In other words, relative is the difference between levels without regard to the power or voltage of either level. Absolute is the power or voltage of a level, traceable to metrology standards. Most measurements in cable systems are relative to the visual carrier. The spectrum analyzer is ideally suited for making relative frequency and amplitude measurements..

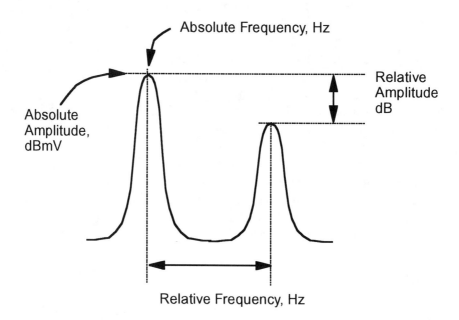

Figure 15. Definition of absolute and relative signal amplitude and frequency measurement.

The Measurement in Brief

Here are the step-by-step measurement procedures of this chapter.

1. Set the spectrum analyzer for a frequency span between 6 and 20 MHz, a resolution bandwidth at 200 or 300 kHz, and use a video bandwidth that is at least as large as the resolution bandwidth.
2. Tune the analyzer to the carrier to be tested.
3. Test for analyzer overload by increasing the attenuator by 10 dB. If the signal level appears to change, leave the attenuator in the highest setting.
4. Use display maximum hold and slower sweep times to assure peak amplitude response.
5. For best accuracy, don't change the input attenuator setting between relative measurements. Bring the carrier peak to the analyzer's reference level. Record the signal level in dBmV.
6. Measure the adjacent aural and visual carriers using relative markers in the same signal span when possible. Use the relative markers to read the signal amplitude differences directly.
7. Use a measurement guardband to remove uncertainty considerations from your system compliance measurements.

Amplitude Measurements

Why Measure Visual Carrier Levels?

Not just because the regulations say so, but because system performance is centered on consistent levels of visual carriers across the full cable system. If a pilot channel, used to set the automatic gain control (AGC) on trunk amplifiers, is either too high or too low, the system flatness can be sent out of whack. If an individual visual signal level is low, it may be modulated by a higher neighboring aural carrier. Low-level visual carriers are especially susceptible to the sound of the adjacent channel interfering with its picture, or if sufficiently low, allow system noise to mask program material with a faded or snowy picture. High signal levels can cause the subscriber's television receiver to overload, producing tearing pictures, and motorboating sounds. Here is a summary of the regulations on absolute visual carrier amplitudes:

Regulation: FCC 76.605 (a)(3)

Regulation text: "The visual signal level, across a terminating impedance which correctly matches the internal impedance of the cable system as viewed from the subscriber terminal, shall not be less than 1 millivolt across an internal impedance of 75 Ω (0 dBmV). Additionally, as measured at the end of a 30 meter cable drop that is connected to the subscriber tap, it shall not be less than 1.41 millivolts across an internal impedance of 75 Ω (+3 dBmV)."

In Other Words: Use a 75 Ω input impedance spectrum analyzer to measure >+3 dBmV at the end of about 100 feet of cable from the subscriber tap.

That seems simple, and it is, once you know how the spectrum analyzer needs to be set-up to measure the visual carrier. To understand that, you need to know a bit about how a spectrum analyzer works.

How a Spectrum Analyzer Measures Amplitude

The spectrum analyzer, as described in Chapter 2, is calibrated to measure the root-mean-square (RMS) voltages of CW signals and present the signals' amplitudes in terms of power. The visual carrier is far from a CW signal. But it does have one characteristic that allows a consistent measure of its power: The modulation is always down from the voltage of the carrier, that is, the level of the voltage peak if it were unmodulated.

The spectrum analyzer is a tuned voltmeter with variable bandwidths and a peak voltage detector as shown in Figure 16. The intermediate frequency (IF) portion processes all the signals mixed down to it from the analyzer's input.

The IF signal is amplified and filtered prior to detection. The filter, called the resolution bandwidth filter, sets the bandwidth of the signal to be detected. For a CW signal, the filterdetermines the shape of the signal displayed on the analyzer's CRT. Like any filter, it needs to be fully charged to display an amplitude that correctly represents the level of the input signal. The narrower the filter, the more time required to charge to the signal's maximum.

Figure 17 shows a CW signal (a) correctly displayed and (b) with considerable amplitude error. In this case, the resolution filter does not charge up because the analyzer's sweep time is set too fast, not allowing the signal to remain in the IF path long enough for thefilter capacitance to accept the full charge. The sweep speed of the analyzer is set

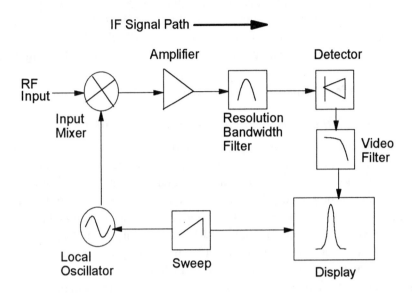

Figure 16. Simple block diagram of a spectrum analyzer.

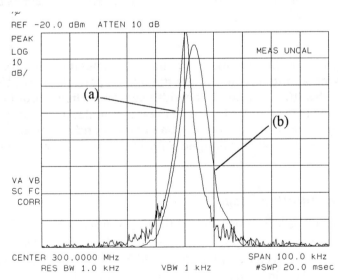

Figure 17. CW signal with (a) normal resolution filter response. The same signal (b) displayed with faster sweep time.

manually to show this effect. Normally the spectrum analyzer automatically controls the bandwidths and sweep speeds to maintain an amplitude calibrated display. Note that the spectrum analyzer has placed a warning on the display, MEAS UNCAL, to indicate that its settings have uncalibrated the amplitude and frequency display for CW signals.

How Filters Affect Signal Appearance

Signals Take the Shape of the Resolution Bandwidth Filter

A spectrum analyzer displays signals in the shape of the resolution bandwidth filter. This is why a CW signal does not appear as a single vertical line at its frequency. A similar effect to that of Figure 17 occurs if the input signal has fast rise and fall times, that are too fast for the analyzer's filter to respond. The television visual carrier, comprised of modulation up and down sharply from the carrier envelope at vertical and horizontal sync pulses, is such a signal. A spectrum analyzer filter that would normally respond to a CW signal with a given preset sweep time may not be able to respond to the visual carrier's modulation. The television signal is composed of fast-changing signal amplitudes. A CW filter does not have the bandwidth to allow the filter to charge to its maximum and stay there for any reasonable sweep time. In other words, when measuring the visual carrier, the spectrum analyzer's resolution bandwidth must be manually set wider than its preset value for measuring CW signals.

TIP

If you find that the spectrum analyzer is acting sluggish, that is, taking a long time to update the display, check the resolution bandwidth and video bandwidth settings. They should be set to 300 kHz or wider for these measurements. In spans wider than 20 MHz, make sure the bandwidths and sweep time are "coupled," that is, changed automatically with frequency span.

Resolution Bandwidth Filter Shape

The resolution bandwidth is a Gaussian-shaped filter, as you see in Figure 17(a). This filter shape is used because its shape allows the fastest sweeps possible without errors. The width of the filter is given as its 3 dB or 6 dB bandwidth points, that is, the distance between the points on either side of the maximum responses which are 3 or 6 dB down from that peak. The 3 dB points are shown in Figure 18. Signals at the analyzer's input are displayed in a shape like the resolution filter. The wider the filter, the wider the signal looks on the display.

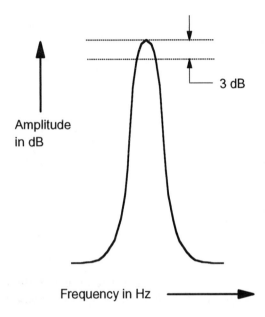

Figure 18. Naming convention for the spectrum analyzer's resolution bandwidth filter.

Use 300 kHz Resolution Bandwidth for Visual Carrier Amplitude

But how wide is wide? The true test to determine if the analyzer's filter is wide enough to fully respond to the input signal is to widen the filter and watch until the amplitude of the signal no longer increases. Figure 19 shows this case for a visual carrier seen with 30 kHz, 300 kHz, and 3 MHz filter settings. The amplitude from the 300 kHz to 3 MHz step did not change amplitude, so it is safe to assume that the 300 kHz filter is wide enough to accept the visual carrier's peak responses as long as the sweep time is set slow enough. Assume for now that the sweep speed the analyzer selects for proper display of a CW signal with a 300 kHz resolution bandwidth is also slow enough to show the visual carrier peak. More is said about the relation between carrier modulation, analyzer sweep speed, and trace display techniques later.

Since the response for the setups between 300 kHz and 3 MHz are unchanged in amplitude, why not always leave the analyzer in its widest resolution bandwidth when measuring the visual carrier? Figure 20 displays shows the same sequence of filter changes, this time for a visual carrier that has an adjacent audio carrier.

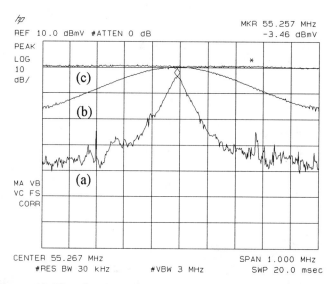

Figure 19. The visual carrier amplitude measured with three resolution bandwidth filters (a) 30 kHz, (b) 300 kHz, and (c) 3 MHz.

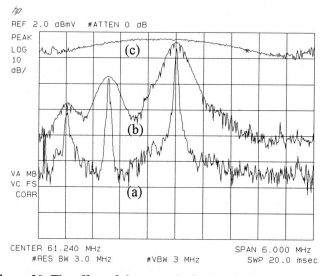

Figure 20. The effect of three resolution bandwidths of the amplitude of the visual carrier (a) 30 kHz, (b) 300 kHz, and (c) 3 MHz.

Another increase in the display amplitude occurs from the 300 kHz to 3 MHz filter step. In Figure 20, the 3 MHz filter adds energy of the adjacent aural carrier to the power of the visual carrier under measurement. Here is how you can tell: In (a) the analyzer is responding to the energy bursts of the visual carrier. In (b) the adjacent aural carrier is still separated from the visual carrier under test by a "valley" of about 20 dB. The visual carrier level, as measured by the marker placed on the maximum signal response, is entirely due to the visual carrier signal. But as the bandwidth is widened, as in (c), the aural carrier energy is included in the measured signal's bandwidth, resulting in another increase in the signal response when the analyzer bandwidth is increased from 300 kHz to 3 MHz.

In summary: Use a resolution bandwidth between 200 and 300 kHz for the measurement of the visual carrier amplitude. Select a sweep speed slow enough to make the trace response smooth.

Example 8. Visual Carrier Level Measurement

From a preset condition, narrow the analyzer's frequency span to 6 MHz, the span of a single channel. Select a visual carrier by selecting the center frequency control and typing in the carrier's frequency. Change the resolution bandwidth to 300 kHz.

Use the marker peak function to get a quick measurement of the visual carrier's level. If this value shifts more than 0.2 dB, use the maximum hold function of the display to capture just the maximum value at each frequency point across the display, and remeasure the amplitude. The use of maximum hold will be clarified later.

Video Bandwidth Filter Must Be Wide Open

One other filter is in the IF path of the spectrum analyzer. The video filter shown in Figure 16 has many uses, the most important for smoothing the effects of random or thermal noise for the detection and measuring of small signals. The video filter is a low-pass filter that may further reduce the bandwidth of the detected and filtered IF signal. This smoothing tends to average the signal response of the heavily modulated visual carrier, making the peak response lower than it really is. Figure 21 shows this effect.

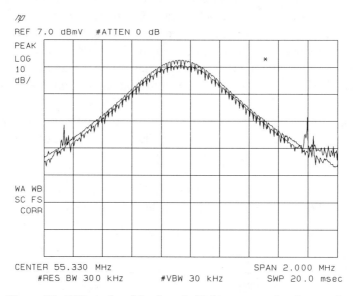

Figure 21. Effect of a video bandwidth set at a value lower than the resolution bandwidth.

The upper response is taken with a video bandwidth of 300 kHz, and the lower response, at a bandwidth of 30 kHz. The 30 kHz video filter averages out some of the visual carrier's modulation, making the level lower and inaccurate. The video filter needs to be out of the way, that is, equal to or wider than the IF's resolution bandwidth for an accurate visual carrier level measurement.

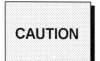

CAUTION   **Resolution bandwidth or video bandwidth set too narrow results in an inaccurate amplitude measurement.**

Visual Carrier Level Measurement Procedure

The carrier measurement with a spectrum analyzer, now that all the potential pitfalls have been exposed, is a very simple procedure. Just remember to set the resolution bandwidth to 300 kHz and the video bandwidth equal to or wider than the resolution bandwidth.

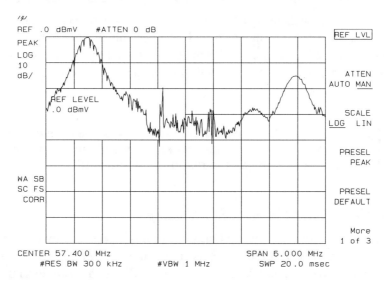

Figure 22. Measurement of the visual carrier level.

Example 9. Measure the visual carrier level.

Tune the analyzer's center frequency to the desired channel. Set the frequency span anywhere from 1 MHz to 6 MHz, keeping the visual carrier on screen. It does not have to be in the center of the display. Set the resolution bandwidth to 300 kHz. Set the video bandwidth anywhere from 300 kHz to 3 MHz to eliminate any of its effects. Use the amplitude controls, to bring the signal to the top graticule of the display. Read the visual carrier's level as the reference level of the analyzer, 0 dBmV, as in Figure 22.

The procedure of keeping the video bandwidth equal or greater than resolution bandwidth can be automated in many spectrum analyzers by setting the analyzer's ratio of video bandwidth to resolution bandwidth to one or greater.

Reading the Visual Carrier Level of Suppressed Sync Scrambled Channels

Even in scrambled channels using horizontal sync suppression, there is sufficient energy in the vertical pulses to see the visual carrier's level. The only difference between sync-suppressed scrambled and normal channel levels is that you have to wait longer for the sync pulses to accumulate at the displayed signal peak. This is done by slowing the sweep speed.

For carrier level measurements, always use a resolution bandwidth greater than 200 kHz and a video bandwidth that is at least as wide as the resolution bandwidth.

Accuracy of the Visual Carrier Level Reading

The accuracy of the visual carrier level measurement is determined by the accuracy of the spectrum analyzer and the test procedure used. Accuracy is evaluated as the uncertainty of a measurement, whether absolute or relative, that is, whether a power level in dBmV or a power differential in dB. The carrier level test is an absolute measurement.

Uncertainty expresses the closeness of the result to the true value. Spectrum analyzer data sheet specifications, its accuracy capability, are the analyzer's measurement uncertainties. This means that calibration procedures are traceable to international measurement standards. In the United States standards for power, voltage, impedance, and etc. are traceable to the National Institute of Science and Technology (NIST). The common practice of test equipment manufacturers is to deliver complete specifications and the procedures to verify the specifications in the spectrum analyzer's operation guide or support manual. Table 2 has data sheet specifications from a typical RF spectrum analyzer relevant to the carrier level test. A full example data sheet is in Appendix C.

Table 2. Spectrum analyzer amplitude specifications.

Maximum Safe Input
  Peak power +72 dBmV (0.2 watts), input attenuation ≥10 dB
  Gain compression >10 MHz ≤0.5 dB (+39 dBmV at input mixer)
Reference Level
  Range: Same as amplitude range
  Resolution: 0.01 dB for log scale, 0.12 % of ref level for linear scale
  Accuracy (referred to +29 dBmV ref level): +49 to −10.9 dBmV, ±(0.3 dB + 0.01 × dB from + 29 dBmV)

Frequency Response
Absolute ±1.5 dB
Relative flatness ±1.0 dB

Calibrator Output
Frequency 300 MHz ±(300 MHz × frequency reference error)
Amplitude +28 dBmV ± 0.4 dB

Input Attenuator
Range: 0 to 70 dB in 10 dB steps
Accuracy: 0 to 60 dB ±0.5 dB at 50 MHz, referenced to 10 dB attenuation setting, 70 dB ±1.2 dB at 50 MHz, referenced to 10 dB attenuation setting

Display Scale Fidelity
Log incremental accuracy: ±0.2 dB/2 dB, 0 to 70 dB from reference level
Log max cumulative: ±0.75 dB, 0 to −60 dB from reference level, ±1.0 dB, 0 to −70 dB from reference level
Linear accuracy: ±3% of reference level

The following examples show how to use the data sheet specifications to determine the accuracy of the carrier level and how procedure affects that accuracy.

Example 10. Accuracy of the carrier level.

A carrier level is measured as +1.2 dBmV by using the procedure in the last example. What is the measurement accuracy assuming the spectrum analyzer specifications are as given in Table 2?

The signal level is measured at the analyzer's reference level; therefore, its accuracy depends solely on reference level accuracy. From Table 2, this given as ±(0.3 dB + 0.01 × dB from + 29 dBmV). Since our level reading was +1.2 dBmV, the accuracy is ±(0.3 + 0.01 × (29 − 1)) = ±(0.3 + 0.28) = ±0.58 dB. This means that the carrier level could be +1.2 dBmV ± 0.58 dB, or, a value between +1.78 dBmV and +0.62 dBmV.

In Example 10 the visual carrier was brought to the top of the screen to measure the signal's amplitude. Why not use the marker to measure the signal amplitude? In some cases the marker reading is less accurate. The following example helps you decide if the marker accuracy is sufficient for your amplitude measurements.

Example 11. Measurement of the visual carrier amplitude with the marker

A visual carrier is measured on the display as shown in Figure 23. Determine the accuracy of the measurement and compare the accuracy with that of Example 10.

The amplitude level is −0.17 dBmV. The reference level is +21 dBmV. The specifications important to this level of accuracy are the reference level accuracy, just as in the last example, and the display scale fidelity. Scale fidelity is added because the signal is at a place other than the reference level. The analyzer's absolute amplitude accuracy is based upon factory calibration of the reference level. Signals not on the reference level have additional uncertainty due to the nonlinearities of the circuits that drive the analyzer's display.

The reference level accuracy is ±(0.3 dB + 0.01 × dB from + 29 dBmV). Since our level reading was −0.17 dBmV, the accuracy is ±(0.3 + 0.01 × (29 −0.17)) = ±(0.3 + .29) = ±0.59 dB, very close to the value in Example 10. The amplitude distance from the reference level is needed to calculate the display scale fidelity. It is simply the

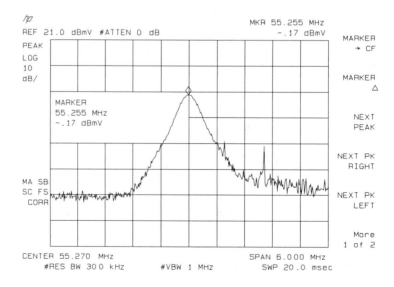

Figure 23. Measurement of the visual carrier with the analyzer's marker.

difference, +21 dBmV – (–0.17 dBmV) = 21 dB. Display scale fidelity is ±(0.2 dB/2 dB) × 21 dB = ±2.1 dB, which is over the uncertainty limit set by the log maximum cumulative specification at ±0.75 dB.

The total uncertainty is the sum of the reference level accuracy and the display scale fidelity accuracy, ±(0.59 + 0.75) = ±1.34 dB. This means that the carrier level is –0.17 dBmV ± 1.34 dB, that is, a value between –1.51 dBmV and +1.17 dBmV.

Although the marker is convenient, its accuracy may prevent use in some instances.

For the carrier level accuracy, bring the carrier peak to the analyzer's reference level.

Guarantee the Accuracy with a Guardband

The compliance carrier level measurement requires that the level be over a certain value at the subscriber tap or terminal. How do you guarantee that the conditions are met with the uncertainty just calculated? The secret is simple. Just add a measurement cushion, or guardband, to the value measured based upon the uncertainty calculated for the spectrum analyzer.

Example 12. Adding a guardband to carrier level measurement.

For Example 8, what measurement guardband will assure that the cable level measurement will be within compliance levels?

The specification for the subscriber terminal is 0 dBmV. But let's say that this tap level cannot fall below +1.0 dBmV. A minimum signal level for the spectrum analyzer needs to be calculated such that as long as the signal level is over this value, the tap level will be within the specification.

The worst case is when the analyzer readout is right at the specification, 0 dBmV in this case. For this readout, the actual input signal level could be 0 dBmV ±0.58 dB, that is, between +0.42 dBmV and +1.58 dBmV. Since the worst case is when the actual signal is 0.58 dB below the analyzer readout, a guardband is added to the anticipated readout: +1.0 dBmV + 0.58 dB, or +1.58 dBmV. As long as the analyzer's readout shows a signal at the reference level of +1.58 dBmV or greater, the signal level will be at or above the desired level of +1.0 dBmV.

To summarize: for a guardband on a visual carrier signal level, add one half the total uncertainty of the readout to the required specification. Any signal at or above the guardband readout value is in compliance.

Overload Causes Major Inaccuracies

Probably the most frequent cause of large measurement errors of amplitude is overload, which occurs when the power at the spectrum analyzer input is so high the analyzer fails to display correct signal amplitudes. The cause is simple. The spectrum analyzer's first mixer operating point is pushed into its nonlinear operating region by the RF input power. The mixer output signals to the IF circuits are compressed, that is, less than they should be, because the mixer can no longer track the power input changes dB per dB. Figure 24 shows this effect. The gain compression point is the total power at the input of the first spectrum analyzer mixer that causes the output response at the analyzer's intermediate frequency, or IF, to be 0.5 dB lower than it would be if the mixer were still operating in its linear region. The linear region is the solid straight line, the nonlinear region is the slope of the dotted line. When the analyzer is operating in compression, the highest input signals may appear lower on the display than they really are, represented by the solid line sloping away from the linear dotted line. The 0.5 dB point is only an arbitrary point picked for specifying analyzer performance. Actual errors can be much greater as the mixer is compressed more and more by increasing input power.

Prior to making any signal level measurement, especially when measuring in a new signal environment such as measuring a tap on a different feeder line, it is good practice to test the spectrum analyzer for overload. Fortunately the test is a simple one, and for carrier level measurements overload is easy to prevent. Figure 25 shows the simple block diagram of the spectrum analyzer with one additional component that can help. Between the input terminal and the first mixer is a variable step attenuator. The attenuator prevents mixer overload by reducing the signal level at the mixer. Since the attenuator is at the analyzer input, it must respond equally to the total frequency response of the spectrum

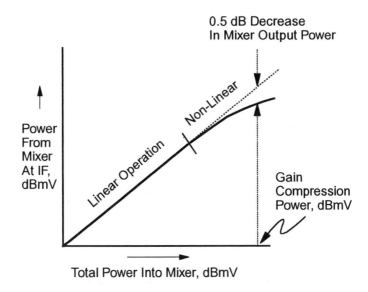

Figure 24. Power in and out of the spectrum analyzer input mixer.

analyzer and cable system. But in doing so, it can reduce the level of all the signal's input at the same time.

TIP

Change the analyzer's input attenuator to determine if the input signals are causing compression overload. The signal level should not change when the attenuation is changed.

The gain compression specification is given as the minimum power at the mixer where 0.5 dB of compression is likely to occur. The specifications in Table 2, under gain compression, specify the example analyzer's total input power at +39 dBmV for a gain compression of less than 0.5 dB. Remember, the total power is at the mixer, not at the input. Only when the attenuator is set to 0 dB are the mixer and input powers equal.

If you even suspect that the analyzer is overloaded, simply increase the analyzer's attenuator setting by 10 dB. If the visual carrier, or any signal for that matter, continues to change amplitude, then the analyzer's input mixer continues to operate in its nonlinear

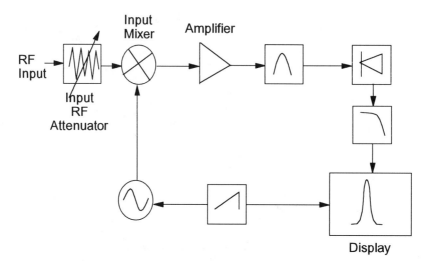

Figure 25. Input attenuator of the spectrum analyzer.

If you even suspect that the analyzer is overloaded, simply increase the analyzer's attenuator setting by 10 dB. If the visual carrier, or any signal for that matter, continues to change amplitude, then the analyzer's input mixer continues to operate in its non-linear mode. Change the attenuator until the signal level stops changing, and leave the attenuator at that setting.

Example 13. Measurement of the visual carrier for analyzer compression (overload).

Set the analyzer for proper measurement of the visual carrier amplitude, that is, a resolution and video bandwidth of 300 kHz. Bring the amplitude of the carrier near the reference level, and shift the amplitude log scale into 1 or 2 dB per division. Mark the amplitude of the trace with a marker, and increase the input attenuator by one 10 dB step.

If the amplitude changes by 0.5 dB or more, as in Figure 26, then you know that the first attenuator setting is not sufficient for overload protection. In this figure the bottom trace was taken with the input attenuator at 0 dB. When the attenuator is switched to 10 dB, the amplitude increase is 0.5 dB, as shown in the top trace. The analyzer was in gain compression with the attenuator set to 0 dB.

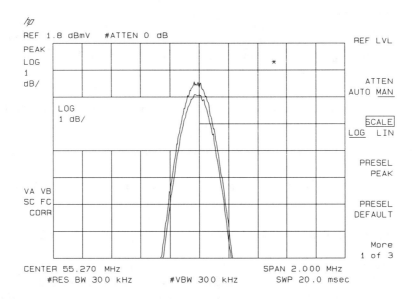

Figure 26. The visual carrier amplitude in compression (bottom trace) and without compression errors (top trace).

Increase the attenuator one more 10 dB step just to ensure that the second setting is sufficient protection for the mixer. If the second and third attenuator settings do not change the signal's amplitude, leave the analyzer in the first change tried, 10 dB in this example.

An attenuator external to the spectrum analyzer can just as easily be used to eliminate the overload conditions, that is, the signals behave as they did when changing the internal input attenuator. The analyzer's amplitude measurements read low by the amount of the external attenuation. When the attenuator inside the analyzer is used, the value of the carrier amplitude is read out for absolute power at the analyzer's input, and no computation is required.

The analyzer traces in Figure 26 were gathered with a display feature known as maximum hold, which keeps only the maximum points for each frequency point as the analyzer sweeps. Maximum hold assures that the trace captures the carrier's peak voltage swings when using a finer amplitude resolution scale of 1 or 2 dB per division.

Gain Compression, the Analytical Approach

Using the analyzer to test itself is good practice, but if you would like to be more analytical about compression overload, a simple formula predicts how much power is at the analyzer's input.

$$\text{Total power} = (\text{power of one channel}) + 10 \log_{10}(\text{number of channels})$$
$$\text{Equation 4}$$

Example 14. Calculate the total power from a system with 62 channels at −4.0 dBmV.

The total power from Equation 4 = −4 dBmV + 10 $\log_{10}$(62) = +13.9 dBmV.

The overload (or gain compression specification) limit for the analyzer is +39 dBmV. The input to the analyzer's input mixer does not cause overload. In fact, the input attenuator could be set to 0 dB, if sufficient care is taken in measuring the signal's accuracy. More will be said about the attenuator's contribution to measurement uncertainty in Chapter 8.

The power at the analyzer's mixer is the total power, as computed, minus the analyzer's attenuation, which is a combination of internal and/or external attenuator settings.

Adjacent Visual and Aural Carriers

The visual and aural carriers of transmitted, or over-the-air, channels are just a few dB different in amplitude. The power difference between aural and visual signal levels within the cable system is increased to avoid adjacent channel interference and intermodulation distortion. Here are the compliance rules:

Regulation: FCC 76.605 (a)(4) and (5)

Regulation text: "The visual signal level on each channel, as measured at the end of a 30 meter cable drop that is connected to the subscriber tap, shall not vary more than 8 decibels within any six-month interval which must include four tests performed in six-hour increments during a 24 hour period in July or August and during a 24 hour period in January or February, and shall be maintained within: (i) *3 decibels (dB) of the visual signal level of any visual carrier within a 6 MHz nominal frequency separation;* (ii) *10 dB of the visual signal level on any other channel on a cable television system* of up to 300 MHz of cable distribution system upper frequency limit, with a 1 dB increase for each additional 100 MHz of cable distribution system upper frequency limit (e.g., 11 dB for a system at 301-400MHz; *12 dB for a system at 401-500 MHz*; etc.); and (iii) a maximum level such that signal degradation due to overload in the subscriber's receiver or terminal does not occur. The RMS voltage of the *aural signal shall be maintained between 10 and 17 decibels below the associated visual signal level.* This requirement must be met both at the subscriber terminal and at the output of the modulating and processing equipment (generally the head end). For subscriber terminals that use equipment which modulate and remodulate the signal (e.g., baseband converters), the RMS voltage of the aural signal shall be maintained *between 6.5 and 17 decibels below the associated visual signal level* at the subscriber terminal." (Italics by the author.)

In Other Words

In one channel

- Aural carrier 10 to 17 dB below visual carrier
- Or 6.5 to 17 dB when the channel is processed by a baseband converter

In system

- <3 dB between adjacent visual carriers
- <12 dB between any other visual carriers
- >10 dB between adjacent aural and visual carriers

These specifications are summarized in Figure 27 except for the 24-hour conditions in part (4). All the measurements are relative amplitudes between carriers. To make all the measurements on an entire system, you could take the absolute level of all the visual and aural carriers, then use math to subtract all the relevant pairs to ensure compliance. The analyzer and a bit of common sense provide a better way.

Aural Carrier Amplitude

To make relative carrier measurements the aural signal must be measured just as accurately as the visual carrier. The aural carrier is an FM signal whose bandwidth is 80 kHz, much less than the visual carrier, so its energy is spread over 80 kHz. Using a

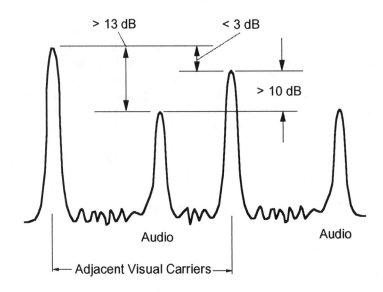

Figure 27. The relative amplitudes of carriers for compliance.

spectrum analyzer bandwidth much wider guarantees that the full aural power is represented by the peak of the analyzer's response. Since 300 kHz resolution bandwidth is used to measure the visual carrier, the same bandwidth is convenient for the aural carrier too. In fact, it is preferred so that both visual and aural carriers can be on the same displayed frequency span, making measurements faster and more accurate. More information on the nature and measurement of the aural carrier will be discussed in Chapter 12.

When making relative carrier measurements, display the two signals in the display at the same time, and use the markers to measure their differences.

Why the Spectrum Analyzer Is Good at Relative Amplitude Measurements

The spectrum analyzer is good at making comparative measurements because of the way signals are displayed: with wide or narrow frequency ranges and wide amplitude ranges simultaneously. Example 15 is a comparison of the visual and aural carriers in one channel.

Example 15. Comparison of the relative amplitudes of the visual carrier and aural carrier in one channel.

Place the channel in the analyzer's display using a 6 MHz frequency span. Set the bandwidth and video bandwidth for an accurate measurement of the visual carrier (300 kHz). Place a marker on the visual carrier peak and a second marker on the aural carrier. Read out the difference. Figure 28 shows a difference of 14.9 dB, well within the 10 (or 6.5) to 17 dB specification.

The accuracy of this measurement is the display scale fidelity between each amplitude and the reference level. Using the specifications in Table 2, with the signals more than 10 dB from the reference level, the uncertainty is ±0.75 dB for each one. The difference between the carriers is 14.9 dB ±(2 × 0.75) dB = 14.9 ±1.5 dB , or between 13.4 and 16.4 dB.

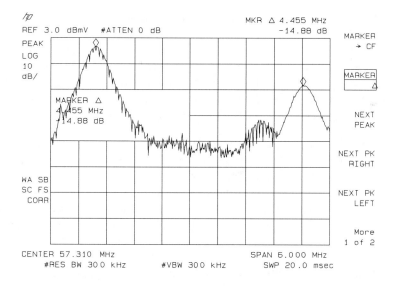

Figure 28. Comparison of visual and aural signal levels.

The uncertainty in Example 15 seems high at ±1.5 dB. It is not too difficult to live with uncertainty of this magnitude when the regulations have such a wide latitude, that is, acceptable ranges of either 6.5 to 17 dB or 10 to 17 dB. But this measurement can easily be made more accurate by bringing the visual carrier to the reference level prior to measuring the amplitude differences.

Bring the uncertainty is the display scale fidelity of only the one signal <u>not</u> at the reference level, and uncertainty of ±0.75 dB. That is, the signal difference of 14.9 dB has an uncertainty of ±0.75 dB. The difference between the visual carrier and the aural carrier is somewhere between 14.15 and 15.65 dB

The accuracy of the analyzer in measuring two signals, which are close in frequency, is very good compared to the demands of this amplitude measurement. But how, in the more exacting measurement of adjacent channel visual carriers, can guardband be applied to meet the more stringent 3 dB minimum specification? Example 16 illustrates.

Example 16. Measure adjacent visual carrier amplitudes.

Use about an 8 MHz span to get both visual carriers on the display. Set the analyzer display amplitude scale from 3 to 5 dB per division, and move the higher carrier to the reference level. Use the maximum hold trace function to assure that the signals' maximum peaks are displayed. Stop the sweep with either the single-sweep mode or trace-view mode, and use the markers to measure the difference, as in Figure 29.

The difference between the carriers is 1.74 dB. Ignore the minus sign in the readout; it only shows that the first marker is placed on the higher signal. To calculate how accurate this number is, refer to Table 2 again. The only specification required is the display scale fidelity. Since the higher signal was placed on the reference level, the only uncertainty is from the second, or lower, signal. The uncertainty is ±0.2 dB/2 dB of amplitude difference, or ±(0.2/2) × 1.74 dB = ±0.174 dB. Round it off to ±0.2 dB.

The adjacent carriers are between 1.74 ± 0.2 dB, or between 1.54 and 1.94 dB apart. Ignore the 1.54 dB value because it is closer to compliance. The worst case is when the carriers are 1.74 + 0.2 dB = 1.94 dB apart in amplitude.

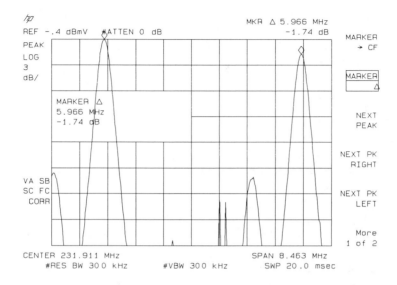

Figure 29. Comparison of the amplitudes of two adjacent visual carriers.

Considering that the analyzer has uncertainty in this measurement, what maximum difference between signals must be measured to assure that the 3 dB minimum regulation is met? As in the past examples of guardband, look at the way the specification is given. One extreme of the uncertainty makes the value better. Ignore that case. But the other extreme would let you accept carriers that were too far apart in amplitude, unless you adjusted the expected analyzer's reading so that every carrier comparison at or better than the adjusted value will meet the 3 dB regulation maximum.

Example 17. Apply a guardband to guarantee adjacent visual carriers.

If the lower signal is at the specification limit, 3 dB down from the other carrier, the spectrum analyzer uncertainty of display scale fidelity says that the actual difference could be worse by the display scale fidelity of $\pm(0.2/2) \times 3.0$ dB = 0.3 dB. If the signal differences were no greater than 3.0 dB − 0.3 dB = 2.7 dB, then the greatest actual signal difference would be 3.0 dB.

In other words, change the specification you test to account for the analyzer's uncertainty. Rather than blow the whistle at 3.0 dB analyzer reading of adjacent carrier levels, consider a difference of 2.7 dB or greater to be just at or out of compliance.

Expect about 1.5 dB uncertainty when measuring relative amplitudes. Because of the way guardband works, your signals are in specification if they appear to be 1.5 dB better than the specification.

Comparing Carrier Levels Wherever They Are

So far, comparisons of carrier levels have been within one or two channel spacings. The measurement procedure changes a bit for the comparison of channel carriers from one end of the system frequency range to the other. Looking back to the carrier level specifications illustrated in Figure 27, visual carriers are the only channel carriers to be compared across wide frequency ranges, since it is their power that dominates the cable system's total energy transmission. Consistency of the carrier levels is the most quickly observed system quality check when the viewer is surfing through the channels. Compliance testing compares aural carriers only to nearby visual carriers.

Using Time to Capture the Levels Accurately

To compare the amplitudes of two signals located far apart in frequency but in the same display, the spectrum analyzer must be capable of resolving the signals with its 300 kHz resolution bandwidth and equal or wider video bandwidth. Figure 30 shows what a 450 MHz-wide system looks like if the bandwidth conditions are met. The roller coaster waves of amplitudes across the band show that the analyzer, in spite of being set to the correct bandwidths, is not displaying the amplitudes well. If the figure's display could come to life, you would see amplitudes of all the signals varying several dB up and down. This happens because is a swept receiver, that is, it can only look at any one frequency point at any one time.

In the analyzer display shown, there are 400 points across a frequency span of 400 MHz, that is, 400 MHz/400 points which reduces to 1 MHz per point. Each point represents a

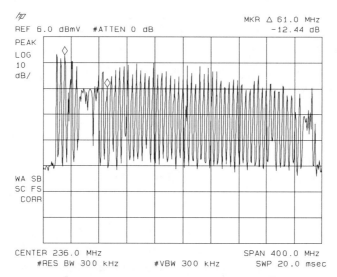

Figure 30. Full system at customer tap with the analyzer incorrectly displaying amplitude.

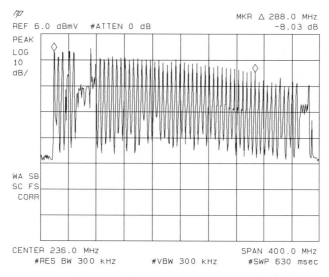

Figure 31. Full system display with sweep time of 630 msec.

single detected analyzer IF response for a time period dictated by the sweep time of the analyzer. In this case, the sweep time is 20 msec, so each point has to be collected, processed, and displayed in 20 msec/400 points, or 50 microseconds (50 x 10^{-6} seconds, also written 50 μsec). The vertical sync pulse, which occurs every 16 msec or so may only occasionally be noticed at that frequency point. The horizontal pulse, which occurs every 63 μsec has a better chance of being captured at each visual carrier point–but not reliably.

To ensure that the correct amplitudes are displayed, lengthen the time at each point so that at least one horizontal sync pulse is always recorded. To allow time for at least ten of the 63 μsec horizontal pulses to occur at each point, the sweep time across the whole span needs to be set to 63 μsec $\times$ 10 $\times$ 400 points, or 630 msec. The results are shown in Figure 31.

The system levels seem more consistent except to the right of the right-most marker, the diamond-shaped symbol riding on top of the signals. These channels are scrambled using suppressed horizontal sync pulse method, so naturally the analyzer cannot show their amplitude unless a less frequent vertical pulse happened along in the 630 msec allotted for each point. So to see all the peaks, use ten repetitions of the vertical sync pulse. The sweep time has to be 16.7 msec $\times$ 10 $\times$ 400, or 67 seconds. Figure 32 shows the results. Now all

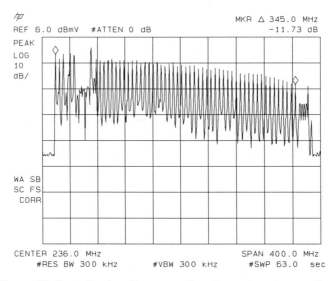

Figure 32. Lengthening the sweep time to capture ten vertical sync pulses at each point.

the channels, including the scrambled channels, are displayed at their full visual carrier levels.

Maximum Hold Captures Carrier Level

An alternative to lengthening the sweep time is to use the maximum hold capability of the analyzer's trace to gather the maximum responses of each value. There is no exact formula to predict how many sweeps bring all the signals to their peak values, but it is easy to see when the signals stop growing. Figure 33 shows the result of using maximum hold with a 20 msec sweep time, collected for about 5 seconds. The results are the same as the slower sweep of Figure 31. You have a choice of which method fits your testing or observation needs, or you can use a combination of longer sweep times and maximum hold.

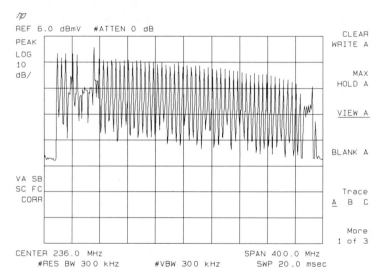

Figure 33. Visual carriers measured with maximum hold over several sweeps.

Maximum hold function of the display trace helps make accurate carrier level measurements faster.

When carriers are compared over smaller frequency spans, the analyzer can be swept faster, and less time is required for the maximum hold method. These techniques are not recommended for regulatory tests because of accuracy considerations. However, broad cable span observations are valuable for examining the general well-being of the system and give a good snapshot of system carrier level quality, especially when environmental conditions change dramatically to affect the performance of field amplifiers and other system hardware.

Relative Carrier Accuracy Revisited

Accuracy and Guardband for Comparing Visual Carriers

When signals are far apart, the spectrum analyzer's own frequency response must be taken into account. Frequency response is the spectrum analyzer's ability to measure a uniformly powered signal over the full frequency range. From Table 2, frequency response, or flatness, for our example analyzer is ±1.0 dB for relative measurements. This uncertainty adds to those already discussed in Examples 16 and 17, making use of the broad span measurements for compliance testing difficult if not impossible. An example illustrates.

CAUTION **Another 1 dB of measurement uncertainty is added when measuring over wide frequency ranges.**

Example 18. Calculate the accuracy of comparing any two signals in a single frequency span, as in the display of Figure 32, and the necessary guardband to ensure meeting the 12 dB regulation.

The amplitude uncertainty for signals within one channel of each other is ±0.75 dB if the higher signal is placed on the reference level, and ±1.5 dB if neither signal is on the reference level. An additional ±1.0 dB is added to each of these for the figure's display if the carriers are more than one channel away from each other.[1] The two testing cases are:

1. If the highest signal is not at the reference level and all others are measured against it, the uncertainty is ±(1.5 +1.0), or ±2.5 dB.

---

[1] This is based upon typical spectrum analyzer frequency response. Flatness changes occur over several tens of MHz, not over one 6 MHz channel spacing.

2. If the highest signal is brought to the reference level, the uncertainty is ±(0.75 +1.0), or ±1.75 dB.

Whichever measurement procedure is used, the appropriate guardband is applied by subtracting the guardband from the specified maximum allowable difference, that is, 12 dB. This ensures that the actual carrier level difference is less than 12 dB. In other words, the difference measured on the analyzer needs to be less than 12 dB − 2.5 dB, or ≤9.5 dB, for procedure 1, and 12 dB − 1.75 dB, or ≤10.25 dB, for procedure 2. For Figure 32 the marker reading shows a difference of 11.7 dB between the highest and lowest visual carriers, indicating a possible regulation violation.

Measuring Carriers One-by-One

Is the measurement accuracy enhanced when the visual carriers are measured individually in narrower frequency spans, then compared? Here is the measurement technique for the best accuracy:

*   Use 300 kHz resolution bandwidth and 300 kHz video bandwidth.
*   Don't change the input attenuator setting.
*   Measure in a frequency span that does not strain the analyzer's ability to capture the peak responses, that is, between 1 to 20 MHz.
*   Measure each visual carrier at the analyzer's reference level.

The uncertainty of using this technique involves comparing the reference level accuracy of each signal and the relative flatness between signals. At first glance, the comparison of two signals, each measured against the reference level, would be a much more accurate comparison than a comparison across a broad span of frequency. This is not always the case because of the way accuracy is specified, that is, the use of ± values.Look at Figure 34. The two signals, I and II, in (a) have an uncertainty indicated by ±A dB and ±B dB, each from some nominal amplitude value. The worst amplitude uncertainties between these two signals are when each is at an opposite extreme, as in (b). In addition, the nominal value around which the uncertainties are specified may not be the same for each signal. In fact, the nominal level may be as much different as the accuracy for each signal, A for signal I and B for signal II. If signal I measures +A when signal II reads −B, or −A for I and +B for II, and the nominal level is also at those extremes, the total cumulative uncertainty in comparing the two signals' amplitudes are shown in (c) as ±(A+B).[2]

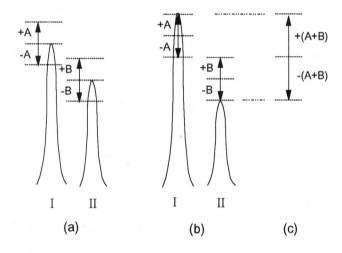

Figure 34. Amplitude accuracy expressed as plus and minus differences from a nominal value.

TIP

For best accuracy, don't change the analyzer's input attenuator after you measure the carrier level.

Example 19. Measure the signals of Figure 32 individually and compute the measurement accuracy.

The visual carrier of channel 2 measures 0.0 dBmV when the signal is brought to the analyzer's reference level with 10 dB of attenuation. The right-most channel in the figure is measured at −11.0 dBmV using the same analyzer bandwidths and attenuator settings. As in Example 3, the measurement uncertainty of both signals due to the accuracy of the reference level is ±0.58 dB. The analyzer's total specified uncertainty for frequency response is ±1.0 dB.

---

[2] This specification interpretation is based upon common spectrum analyzer testing practices.

Since each signal amplitude is known to within ±0.58 dB, the total uncertainty between them is twice that amount, or ±2 × (0.58) dB = ±1.16 dB. The total uncertainty is this plus the analyzer's relative frequency response of ±1.0 dB, for a total of ±2.16 dB.

The measurement guardband, computed here as demonstrated in Example 18, is 12 dB (the FCC regulation for visual carrier level differential) minus 2.16 dB, or 9.84 dB. For this example the difference between the signals is 11 dB, suggesting that the two signals are out of regulation by 11 − 9.84 = 1.16 dB.

The techniques exemplified by Examples 18 and 19 show little difference between test methods. The one best for your compliance or troubleshooting testing depends upon the specifications of your spectrum analyzer.

Select Worst Case Signals to Test

Because there are so many channel carriers to be tested against so many other carriers, the process could become tedious. Automatic testing of these levels is possible using an external or internal computer to control the analyzer, gather the carrier level information, apply the uncertainty guardbands, and display the results in a tabular report. This data can be used along with the results from the procedures in Chapter 5 to satisfy the compliance measurement needs. To make measurement accuracy computation easier, a worksheet and measurement procedure and example are included in Appendix C.

Summary of Carrier Amplitude Measurements

Carrier amplitude measurements are absolute and relative. Absolute means that the power level is known to within a predictable uncertainty. Relative means that the power differences are known to within a predictable uncertainty. Although absolute carrier levels are important, the majority of cable television amplitude measurements are relative.

To make consistently accurate amplitude measurements:

- Run the analyzer's calibration routines periodically, or when the ambient temperature changes more than 5°C.
- Check for, and prevent overload before making amplitude measurements.
- Use 200 to 300 kHz resolution bandwidths.

- Use a video bandwidth that is at least as large as the resolution bandwidth.
- Use display maximum hold and slower sweep times to assure proper amplitude response.
- Measure absolute signal levels at the reference level when possible.
- Measure relative signals in the same signal span when possible, and place the highest at the reference level.

The spectrum analyzer makes carrier amplitude compliance measurements, provided it has absolute and relative performance specifications. To assure amplitude compliance, analyzer accuracy must be computed and included in measurement guardbands. The poorer an analyzer amplitude accuracy specification, the wider the accuracy guardband needs to be.

Review the procedures and examples of this chapter, preferably with your spectrum analyzer in front of you. They help prevent common measurement mistakes and give simple guidelines and short cuts to make your job easier. The lessons learned about spectrum analyzer technique and accuracy are applied throughout the remainder of this book.

Selected Bibliography

- Benson, K. Blair, and Whitaker, Jerry. *Television Engineering Handbook*. Rev. ed., McGraw-Hill, Inc., 1992.
- *Cable Television System Measurement Handbook*. Hewlett-Packard Company, Literature No. 5952-9228, Santa Rosa, CA, January 1977.
- Engelson, Morris. *Modern Spectrum Analyzer Measurements*. Portland: published by JMS, 1991.
- *Code of Federal Regulations, Title 47, Telecommunications, Part 76, Cable Television Service*. Federal Commission Rules and Regulations, 1990.
- Peterson, Blake. *Spectrum Analysis Basics*. Hewlett-Packard Company, Application Note AN 150, Literature No. 5952-0292, Santa Rosa CA, 1989.

System and In-Band Frequency Response

5

Overview

System and in-band frequency response measures the cable system's continuity and individual channel video quality by comparing the relative levels of signals and sidebands put into the system with those along the distribution path. The system frequency response tests give you information about the health of your plant. For example, response suckouts and standing wave patterns indicate electrical or mechanical faults. A complete system sweep as done before a system goes on-line, is not practical after the system is operational because it causes too much interference to the subscriber. Long gone are the late nights when most channels would shut down so you could do your testing. However, several techniques have evolved to make system frequency response measurement efficient and complete while minimizing subscriber disturbance.

Frequency Response Measurements

In-band measurements are designed primarily as bench-top tests for processors and modulators, although most of their on-line responses can be extracted from certain system response test data. This chapter surveys the tests with a spectrum analyzer.

The spectrum analyzer system frequency response measurement trades convenience for guaranteed noninterference with your subscriber for system frequency response. You need to take the analyzer in the field after taking a reference measurement at the head end. The analyzer is well suited for bench and noninterfering in-band measurements, although not all of these are covered in this chapter.

FCC Regulations

There is no specific FCC regulation that demands the system frequency response test. However, the relative visual carrier levels specified in Chapter 4 demand that the system flatness be within certain limit. As you will see, good flatness does not guarantee that the visual carriers levels are within these limits.

> **Regulation**: from FCC 76.605 (a)(4) and (5)
>
> **Regulation text**: "The visual signal level on each channel, as measured at the end of a 30 meter cable drop that is connected to the subscriber tap, shall not vary more than 8 decibels within any six-month interval which must include four tests
>
> performed in six-hour increments during a 24 hour period in July or August and during a 24 hour period in January or February, and shall be maintained within: (i) 3 decibels (dB) of the visual signal level of any visual carrier within a 6 MHz nominal frequency separation; (ii) 10 dB of the visual signal level on any other channel on a cable television system of up to 300 MHz of cable distribution system upper frequency limit, with a 1 dB increase for each additional 100 MHz of cable distribution system upper frequency limit (e. g., 11 dB for a system at 301-400 MHz; 12 dB for a system at 401-500 MHz; etc.); and (iii) a maximum level such that signal degradation due to overload in the subscriber's receiver or terminal does not occur."

Frequency Response Measurements in Brief

For Full System Frequency Response

1. Connect the reference cable signal.
2. Set the span of the analyzer to cover the bandwidth to test, e.g., 50 to 400 MHz.
3. Set the resolution bandwidth as wide as it will go (about 4 MHz in some analyzers).
4. Set the video bandwidth to 300 kHz and sweep time to 8 seconds.
5. Save this trace and analyzer settings in nonvolatile memory.
6. Connect the cable signal to be compared.
7. Recall the reference trace and normalize against the input trace in a 2 dB per division amplitude display.
8. Use the markers to measure the peak-to-valley amplitude range.
9. Calculate the peak-to-peak flatness as one half the peak-to-valley value in decibels.

For In-Band Frequency Response

1. Disconnect or isolate the processor or modulator from your system.
2. Provide a video sweep test signal to the RF input, and connect the spectrum analyzer to the output of the processor or modulator.
3. Set the analyzer on the RF channel frequency with span of 6 MHz, and the resolution and video bandwidths to 300 kHz.
4. Slow the sweep to 200 ms and use the markers to measure the widest signal amplitude variation in the flat valley between the visual carrier and the audio carrier.
5. Divide this value by two to get the ±dB in-band frequency response.

Regulation: FCC 76.605 (a)(6)

Regulation text: "The amplitude characteristic shall be within a range of ±2 decibels from 0.75 to 5.0 MHz above the lower boundary frequency of the cable television channel, referenced to the average of the highest and lowest amplitudes within these frequency boundaries. (i) Prior to December 30, 1999, the amplitude characteristic may be measured after a subscriber tap and before a converter that is provided and maintained by the cable operator. (ii) As of December 30, 1999, the amplitude characteristic shall be measured at the subscriber terminal."

In Other Words

Full system

- ◆ <3 dB between adjacent visual carriers
- ◆ <12 dB between any other visual carriers
- ◆ >10 dB between adjacent aural and visual carriers

In-Band

- ◆ ±2.0 dB over the video of the channel

There are well-defined practices for interpolating the results of a full system frequency response to meet these specifications. However, the spectrum analyzer test results do not provide enough fine-grain system flatness information for this analysis. Only the visual carrier amplitudes are compared with the spectrum analyzer frequency response test in this chapter.

 The spectrum analyzer "visual carrier reference system" frequency response test procedure does not give you sufficient data to measure the adjacent aural and visual carrier levels nor the in-band video frequency response levels.

When System Flatness Goes Bad

System frequency response is a bellwether indicator of system health for preventive maintenance as well as an aid to meeting in-band regulations. Visual carriers set to within ±12 dB of each other at the head end or a hub must maintain that amplitude relationship through cable roll-off and slope compensation in trunk and feeder amplifiers. And, although there is no FCC specification aimed at system flatness, poor noise and distortion performance, which are regulated, are at risk when the system flatness goes askew. When the in-band response goes outside the ±2 dB flatness specification, your subscribers' picture becomes faded and subject to changing and distorted color.

Suckouts and Standing Waves

A more serious problem is a suckout due to a cable or amplifier malfunction. A suckout is a complete loss of cable frequency response as shown in Figure 35. Like a black hole, a suckout removes all signals in its band, as if a band stop filter were placed in the main cable. Incorrectly installed passive components and amplifier performance problems cause suckouts, so it is relatively easy to test taps up the system to find the cause. Suckout complaints are that an entire channel or set of channels is missing from their television reception.

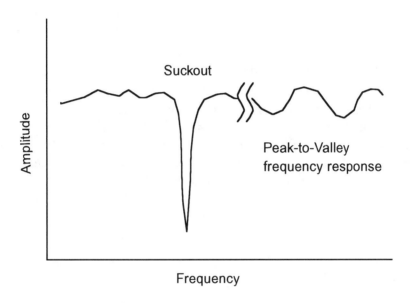

Figure 35. Illustration of standing waves and a suckout.

Another indication of poor installation or a cable fault in the making is the so-called standing wave patterns uncovered by full system frequency response testing. In Figure 35, the periodic response on the right side of the graph is a periodic amplitude variation in the system caused by cable signals being reflected back from a cable or component discontinuity. Common causes are corroded connectors or broken cables on the trunk or feeder lines. The reflections re-reflect on the outputs of other system components, such as splitters and amplifiers, adding and subtracting from the forward traveling signals, causing the periodic signal amplitude variation called standing waves. There are guidelines for acceptable variations of standing waves so you know when they are indicative of system problems, and when they are acceptable. These will be discussed later.

Look for changes in standing waves in frequency response when system capacity has changed.

What is the relationship of standing waves caused by mismatches to the frequency response of the system? Standing waves affect frequency response. But the primary contributor to frequency response is gain over the frequency of the system amplifiers.

In-Band Frequency Response Indicators

Another frequency response measure tests the response in each channel. This is the in-band frequency response. Figure 36 shows the FCC in-band frequency response criteria. The video range measured, from 0.75 MHz above the lower channel boundary to 1 MHz below the upper channel boundary, includes the visual carrier and color burst. Often the frequency response misbehaves as it rolls off toward the higher side of the channel, reducing the effectiveness of the color information. Your subscriber may complain about poor color when this happens.

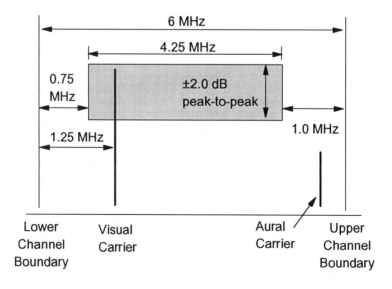

Figure 36. In-band frequency response specification.

Subscriber complaints of missing channels, poor color or fading of a channel usually indicate frequency response problems. With missing channels, look for a suckout.

System Sweep and Frequency Response Techniques

To measure the full system response of a system, you would have to provide a signal at each and every frequency and record the response at each frequency. This is the fundamental component or system measurement technique known as network analysis. Figure 37 shows a diagram of network and spectrum analysis. In network analysis, the device or system under test is positioned between the signal source and the receiver to collect the data. By comparing the known signal at the input of the device under test to the results of the receiver measurement, the device can be characterized over its entire frequency band.

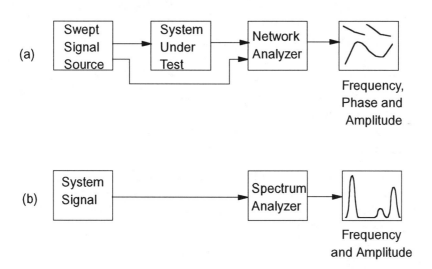

Figure 37. Test techniques of (a) network analysis and (b) spectrum analysis.

System sweep techniques emulate a network analyzer with some clever innovations required because, unlike the network analyzer bench component test, system sweep requires that the source and receiver be far removed from each other. Here is a brief survey of the popular types of system frequency response tests.

High-Level Simultaneous Sweeping

The source is set to an amplitude 15 to 20 dB above the video carrier level and swept over the entire frequency range. The receiver has a broad band detector so that all high-level responses are picked up at all times. The source also sends a coordinated time and sweep speed message out on an unoccupied frequency of the system. The receiver then knows where in its frequency display to place the amplitude response it receives. The high-level signal is swept for 0.2 ms every 5 seconds to keep subscriber interference to a minimum. Precautions against AGC disruption are made by placing band-stop filters at the AGC pilot frequencies at the source output. The disadvantage to this high-level sweep is that it can disrupt digital programming and servicing, such as set-to converter programming.

Low-Level Synchronous Sweep

Similar to high-level sweep, the source is set 30 dB below the visual carrier to minimize disruption of most system signals. A source pilot signal tells a more sensitive, frequency-tuned receiver where to look for the source signal.

Other technologies include a hybrid of high-level sweep and reference sweep, and the most sophisticated test set–the vertical interval sweep system. These systems provide fast, detailed system response information but are often expensive.

Visual Carrier Reference System

The test techniques mentioned so far offer complete system frequency-response data with few disadvantages except expense and the need for coordinated effort by two or more system personnel during the tests. The spectrum analyzer offers convenience at the expense of complete test data.

Although the carrier reference technique does not provide as complete test results as the other techniques, it is convenient and simple when you need the analyzer in the field anyway.

Rather than the analyzer looking for new signals from the hub or head end, it uses the visual carriers themselves for reference and test signals. Of course, this means that the analyzer itself must be carried to the hub or head end, and then to the field for the comparison tests. But when you are taking analyzer in the field, it is easy to add the frequency-response testing to the routine preventative maintenance without a lot of preparation or help from other people.

Here is how it is used. The spectrum analyzer takes a snapshot of all the carriers at the head end or other reference point. The analyzer and its reference test data are taken to test points where the reference and local spectrums are compared using a technique known as normalization.

There are three disadvantages to the reference technique:

- Accounts for the flatness of points 6 MHz apart. No channel flatness can be extrapolated from the data.
- Does not cover flatness of bands where no signals are transmitted. However, dummy channel carriers can be injected for measurement of unoccupied bands.
- Cable signal sources must be stable between the time the reference signals are recorded and the comparison is made. Time varying signals, such as time-variable scrambling or changes in the off-air channels due to weather will not be measured correctly.

TIP

Studies indicate good correlation between all system frequency response test techniques mentioned in this chapter.

Visual Carrier Reference Measurement

The key to accurate system frequency-response measurements with the spectrum analyzer is the understanding of its normalization function.

Normalization

In comparing reference and local signals, a technique called normalization is used. Normalization is simply the graphical subtraction of one response, the reference, from another, usually the signal at the input of the analyzer. Figure 38 shows this graphically. The amplitude response displayed on the analyzer is a series of values plotted on the display that form the shape signals or other responses, as shown in the Figure as (a) and (b). Response (a) represents three visual carriers at different levels, and (b) represents the same carriers recorded at a different time and place. The change of levels in the signals represents the system flatness effects. In (c) the two displays are shown on the same graph. Normalization is the process of subtracting the amplitudes of (a) and (b), and placing the difference amplitude as if it were a trace on the display. If the reference and test responses are equal, the trace would be a straight line, called the normalization line. Trace (d) shows the trace when the responses are different. The higher carrier appears above the normalization line, and the lower response is below. Now let's look at how this is used in the spectrum analyzer to make frequency response measurements.

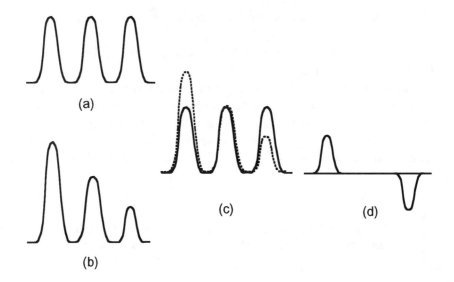

(a)

(b)

(c)

(d)

Figure 38. The process of normalization with (a) the input signal, (b) the reference signal, (c) responses (a) and (b) superimposed on one another, and (d) the difference between (a) and (b).

Example 20. Make a reference sweep measurement for the full system.

Connect the reference cable signal to the input of the analyzer. Set the start and stop frequencies to include all the system channels and pilots. Make sure that the analyzer

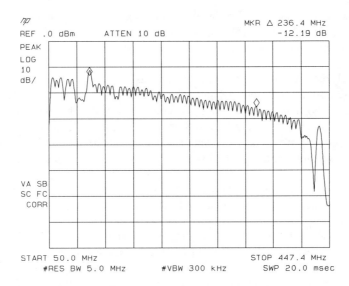

Figure 39. Reference sweep for full system frequency response measurement.

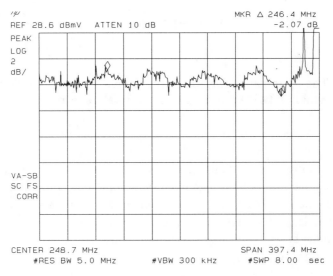

Figure 40. The normalization of the input and reference traces in 2 dB/division.

is not overloaded, especially for signal compression. This is done by changing the attenuator up and down in value and watching for amplitude variations. If none occur, set the attenuator for the lower setting.

Set the analyzer resolution bandwidth as wide as it will go in order to smooth the response. On some analyzers a bandwidth higher than 3 MHz can be set, even though the analyzer is specified for up to only 3 MHz. Set the video bandwidth to 300 kHz and sweep over the range in 8 seconds. Figure 39 shows this sweep. For reference, markers are set at a leakage test signal on the left, and on the highest unscrambled channel on the right. A system test pilot is the response on the far right.

Save this trace in nonvolatile memory. Most modern analyzers provide internal trace storage or a memory card where data can be stored.

Move to another place in your plant, carrying the analyzer and the reference data. Repeat the measurement. Be careful to use the same attenuator setting used in the reference. Recall the reference trace into an analyzer trace memory which can be used with normalization. Subtract the two traces and display the difference in a log trace display of 2 dB/division as shown in Figure 40.

The markers show a variation of about 2 dB peak-to-valley. The peak-to-peak value is one-half this, or ±1 dB.

TIP	**The spectrum analyzer markers give the peak-to-valley reading, which is twice the ± dB peak-to-peak reading.**

The wave pattern in Figure 40 is characteristic of standing waves, indicating a mismatch of impedance along the branch. The high responses on the far right indicate a lack of calibration at these points. The spike near the right is caused by modulation on the pilot tone. Unstable signals, such as time-variable scrambled channels, cannot be used for this measurement. On the far right, the high response is caused by an absence of signal both in the reference and in the test bands.

If your analyzer does not have the ability to save and recall a trace, this procedure is not available to you.

A stable reference signal is required every 6 MHz for the measurement of frequency response using the reference method.

With other sweep methods the slope of the line in Figure 40 can be used to predict the amplitude variation within one channel, therefore helping to determine the in-band flatness as defined in Figure 36. However, since the analyzer reference method uses only the visual carriers for reference and comparison, you cannot find in-band slope with this data.

Extrapolating the in-band slope with the analyzer reference method is meaningless since the only points measured are the visual carrier peaks.

Peak-to-Valley Response Predicts Branch Continuity

How do you know when the peak-to-valley response level is an indication of trouble? There are guidelines for the anticipated levels of standing waves generated by cascaded amplifiers. If the number of cascaded amplifiers is N then you can expect peak-to-valley response of $(N/10 + 2)$ dB, or $(N/10 + 3)$ dB with the addition of a feeder amplifier. In the example, if the tap measured was 14 amplifiers down the branch, the anticipated peak-to-valley standing wave is $(14/10 + 2)$, or 3.4 dB. This is a peak-to-peak value of 3.4/2 or ±1.7 dB. So our example, with a peak-to-peak variation of ±1 dB, is better than anticipated. Therefore, you would not expect trouble brewing.

Guidelines tell you if a standing wave is an indication of trouble.

Comparing Frequency Response with Relative Carrier Levels

How does frequency response relate to the regulation of carrier levels? Carriers that are more than 12 dB apart in amplitude put the system out of compliance. But be careful! The frequency-response test does not tell you about the relative carrier levels. It compares performance between two points in the plant, not two points in the spectrum. The signals could very well be out of specification in both places in the plant and still "pass" the flatness test! However, if the frequency-response slope varies widely, signal-level problems are close to being out of specification, and could be pushed out of compliance elsewhere in the system.

> **TIP** **Good frequency response does not guarantee visual carrier amplitude compliance.**

In-Band Frequency Response Measurement

Except for some new noninterfering test setups, testing modulators and processors on-line for in-band frequency response disrupts service to your customers. The prescribed method of testing is to take the modulator or processor out of service and follow a setup as shown in Figure 41.

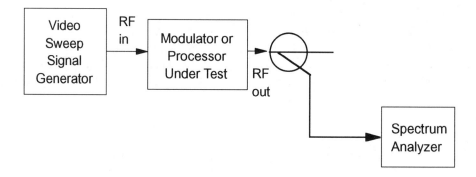

Figure 41. Recommended setup for testing in-band response of head end equipment.

The video-sweep signal generator puts the same test signal on every horizontal line in both fields. Thus, it is called a full-field test signal. The analyzer is tuned to the processor or

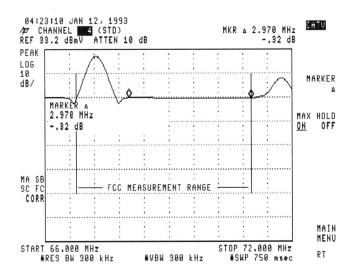

Figure 42. Analyzer response to a full-field sweep.

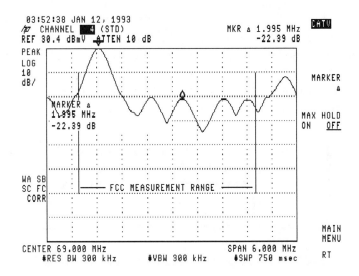

Figure 43. Analyzer response to a full-field multiburst test signal.

modulator test channel output frequency, just as it is for the cable testing. The video trace is collected with the analyzer in its maximum hold display mode to paint out the processor's video response. Then markers are used to measure the video frequency response flatness. Figures 42 and 43 show two such displays.

TIP

To measure in-band frequency response, set the analyzer for maximum hold, 300 kHz video, and resolution bandwidths in a 6 MHz span.

The test signal in Figure 42 is a sweep of the video over the full channel range. The visual and aural carriers poke through the video signal which is set to a level to simulate a black response over the full field. In-band flatness is measured between the markers shown in the figure. The dips on either side of the visual carrier response are caused by the analyzer's detection algorithm, and are not part of the processor's response.

The test signal in Figure 43 is called an FCC multiburst. Multiburst provides video test responses at several frequencies offset from the visual carrier. Flatness is a measure of the peak response of these sidebands relative to one another. The analyzer's bandwidth is set to 300 kHz in order to capture the full signal responses of the higher frequency test bursts with the penalty that the 0.5 MHz signal is buried under the visual carrier response. However, in typical modulator and processor alignment adjustments, it is usually the higher frequency offsets that roll off, not the close-in response. The analyzer's resolution bandwidth can be set narrower to resolve the 0.5 MHz multiburst, at the risk of uncalibrating the high frequency bursts.

CAUTION

The spectrum analyzer 300 kHz resolution bandwidth does not resolve the 0.5 MHz burst. Narrower bandwidths don't let you read the high-frequency bursts accurately.

Example 21. Measure the in-band frequency response using full-field sideband sweep.

Set up your modulator as shown in Figure 41 with the modulator either out of service or set to a channel that has been set up for test purposes. Set the video generator to provide a full-field sweep. Set the spectrum analyzer to the channel frequency output from the modulator with a span of 6 MHz. Set the resolution and video bandwidths to 300 kHz, the sweep speed to 750 ms, and the trace to maximum hold. After the trace has stopped changing amplitude, which should be is just a few seconds, place the markers at the minimum and maximum points on the flat space between the dip on the right of the visual carrier, and where the audio carrier begins to rise out of the sideband response. The measured value in the figure is 0.32 dB, the peak-to-valley value. The specification for in-band flatness is the ± variation, so it is one-half the peak-to-valley value, or ±0.16 dB peak-to-peak.

TIP

The two analyzer markers read in-band flatness as a peak-to-valley value. For comparison to the ±2.0 dB compliance level, divide the peak-to-valley value in half.

Sources of Measurement Uncertainty

Analyzer frequency response and amplitude-scale fidelity specifications affect the accuracy of system and in-band frequency response.

If the same analyzer is used in the reference and field test for the system frequency response tests, the analyzer's frequency response is normalized out of the measurement. One exception to this is if the analyzer's input attentuator is changed between the reference and the field tests. Analyzer frequency response is usually specified only for the 10 dB attenuator setting. Use the same attenuator setting for the reference and field tests.

For accuracy and consistency, you should use the same analyzer for the system frequency-response measurement. If different analyzers are used, the penalty is at least twice the relative frequency response, about ±2.0 dB for the examples. Slight frequency offsets between analyzers can also cause uncertainty in the form of trace ripples set 6 MHz apart.

Display-scale fidelity adds about ±0.75 dB over the amplitude ranges typically measured. Double this value if different analyzers are used to make the reference and field measurements. You can ignore frequency response for the in-band measurement because of the comparatively narrow frequency span. Display-scale fidelity accuracy depends upon the amplitude differences of the markers. Here is an example.

Example 22. Calculate the accuracy for an in-band flatness measurement.

The analyzer markers show a peak-to-peak in-band flatness of 3.4 dB. What is the accuracy of the measurement?

The in-band flatness is one half the peak-to-peak value, or ±1.7 dB. The display scale fidelity specified is ±0.2 for every 2 dB. The in-band flatness has an uncertainty of about ±2 dB, so the flatness is between ±1.5 and ±1.9 dB.

Choose a test point where the system levels are as close to your reference measurements as possible, that is, a place with unity gain. This way slope or gain changes do not interfere with the test results. If the reference signals are close in amplitude to the reference, the temptation to use a different attenuator setting will not be a problem.

Summary

This chapter covers the two cable television system frequency-response measurements: system frequency response and in-band frequency response. Several system sweep techniques are available.

System Frequency-Response Measurement

The spectrum analyzer provides a simple, noninterfering technique for measuring system frequency-response, although it does not provide flatness data where there are no signals. Its response trace cannot be used to extrapolate in-band or adjacent visual and aural level changes.

For the analyzer system frequency-response measurements:

- Set the analyzer for as wide a resolution bandwidth as possible and set the video bandwidth to 300 kHz and the sweep time to 8 seconds.
- Save the trace in nonvolatile memory as the reference.
- In the field, use the same analyzer with the same input attenuator setting.
- Subtract the reference display from the input display, and show it in 2 dB/ division.

Here is what can be learned from the response:

- Standing wave responses indicate a system continuity problem.
- Suckouts may be a component failure.
- Confirmation of unity gain and proper slope application.

In-Band Frequency Response

The analyzer is ideal for viewing and measuring the video response of a modulator or processor.

Set the analyzer for:

- Span of 6 MHz.
- Resolution and video bandwidths of 300 kHz.
- Trace maximum hold.
- Use two markers to measure the peak-to-peak flatness.

The in-band frequency response is one-half the peak-to-peak reading.

Selected Bibliography

- Daugherty, Mark. *Sweep Technology Comparisons.* Communications Technology Publications Corp., November 1990.
- *HP 85721A Cable TV Measurements and System Monitor Personality.* Hewlett-Packard Company, User's Guide, Part No. 85721-90001, December 1993.
- *NCTA Recommended Practices For Measurements on Cable Television Systems.* 2nd. ed. October 1993.
- Windle, Steve. "System Sweeping", *Communications Engineering and Design,* June 1989.

Measuring Carrier and Sideband Frequencies

Overview

The measurement of frequencies are the simplest measurements made, provided you have capable equipment. As in Chapter 4, the lessons learned in these frequency measurements teach a great deal about the capability and limitations of the spectrum analyzer. This chapter teaches you how to make frequency measurements on signals that look too well-modulated to measure.

Frequency Measurements

Although frequency and amplitude measurements are treated separately in this book, in practice they are made with similar setups and at the same time. The remainder of this chapter discusses relative and absolute frequency measurements, when to test, accuracy, and guardbands.

Frequency Measurements on Channel Carriers

The importance of frequency measurements is less critical than those for amplitude because much of the cable system's carrier frequencies are controlled by head end heterodyne processors and demodulators-modulators that have high frequency accuracy and stability. The cable distribution may be able to change the amplitudes of signals, but unless there is reallocation of the channel frequencies, the carrier frequencies will remain the same as they are at the head end. Figure 44 shows the frequency conversion paths often used at the head end.

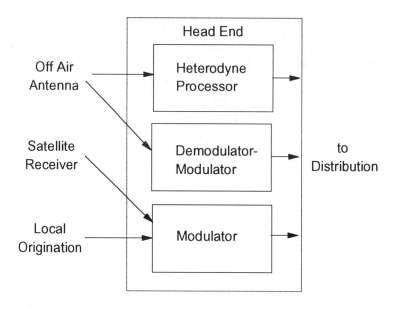

Figure 44. Frequency conversion at the head end.

Frequency Measurement in Brief

If your analyzer has a counter and high-frequency accuracy:

1. Center the channel visual carrier in a span of 6 MHz with the resolution bandwidth set to 300 kHz.
2. Place the marker at the peak of the carrier and turn on the marker counter. The readout is accurate to ±400 Hz if your analyzer's aging rate is 10^{-7} per year and the counter jumps 100 Hz now and then.
3. To measure the audio carrier relative to the visual carrier, place the counter marker at the visual carrier, and the second marker at the audio carrier. The accuracy is ±600 Hz.

If your analyzer does not have a counter marker:

1. Place the carrier near the center of the display, span down to 100 kHz, and place a marker at the carrier or place the carrier at the center frequency. The readout is accurate to about ±4 kHz, that is, a little over 3% of the span. Record this value
2. To measure the audio carrier relative to the visual carrier, move the audio carrier to the center of the display and read the frequency as in the step above. You will have to wait for a quiet moment to catch the audio carrier with the marker to peak function. Subtract the video and aural frequencies. The accuracy is about ±8 kHz, or about twice the single frequency accuracy.

Even though an off-air signal is kept at the same channel within the cable system, it is often demodulated from the antenna source before remodulation and distribution to more closely monitor the signal quality. Other frequency measurements, such as FM deviation, and identification of ingress and distortion products are treated in other chapters.

Absolute and Relative Frequency Measurements

Absolute frequencies of the channels within a cable system are set according to the system's frequency allocation, which is, in turn, dependent upon the system type, number of services, and the total number of channels. The best test for frequency compliance is a satisfied subscriber, because collectively, subscriber TV receivers and/or the set-top converters they rent from you are the best judges of proper frequency settings. At the sign of any drift, reallocation, or errant signals, the business office phones start ringing.

Chances are, you won't have to make many carrier frequency measurements. If the channels are where they belong at the head end, they aren't going to change frequencies in the distribution system unless they go through another conversion stage.

Visual carriers are set to absolute frequencies. Audio carriers are placed relative to their associated visual carrier to maintain subscriber satisfaction. Just as in amplitude measurements, frequency testing is done both absolutely and relatively. The carrier frequencies are measured absolutely, and the audio carriers are measured relatively similar to the way amplitudes are depicted in Chapter 4, Figure 15.

Frequency Allocation Standards

In North America three cable frequency-allocation schemes are used: Standard, HRC, and ICC/IRC. The Standard allocation, which is the most common scheme, is similar to the broadcast allocation, with all the visual carriers except channels 5 and 6 placed 1.25 MHz above 6 MHz multiples. Channels 5 and 6 are placed 0.75 MHz below 6 MHz multiples. HRC stands for harmonically related carriers, where the visual carriers are placed exactly upon the 6 MHz multiples, starting with channel 2 at 54.0 MHz. ICC/IRC, stands for incremental coherent carriers/incremental related carriers. This allocation is the same as the Standard allocation with channels 5 and 6 set 1.25 MHz above the 6 MHz multiples. Table 3 shows the visual carrier frequencies for all three frequency allocations up to channel 13. Both HRC and ICC/IRC reduce composite triple-beat distortion, the subject of Chapter 8. Because the Standard frequency allocation is most common, Standard frequencies are used for the examples in this book.

When Frequency Measurements Are Important

To keep order in the worldwide communication spectrum, transmission frequency standards have been established by the International Radio Consultative Committee, known as CCIR for their French-language title. If all transmissionswere perfect, little frequency measurement would be necessary. Cable systems, whose transmissions are not supposed to be radiated through the air, are tightlyregulated at frequencies used by air navigation equipment. The FCC sets tight frequency standards, so that even if the signals

Table 3. Example visual carrier frequencies for Standard, HRC, and IRC frequency allocations.

CATV Channel	Standard		HRC		IRC	
	Picture	Sound	Picture	Sound	Picture	Sound
2	55.25	59.75	54	58.5	55.25	59.75
3	61.25	65.75	60	64.5	61.25	65.75
4	67.25	71.75	66	70.5	67.25	71.75
-	-	-	72	76.5	73.25	77.75
5	77.25	81.75	78	82.5	79.25	83.75
6	83.25	87.75	84	86.5	85.25	89.75
7	175.25	179.75	174	178.5	175.25	179.75
8	181.25	185.75	180	184.5	181.25	185.75
9	187.25	191.75	186	190.5	187.25	191.75
10	193.25	197.75	192	196.5	193.25	197.75
11	199.25	203.75	198	202.5	199.25	203.75
12	205.25	209.75	204	208.5	205.25	209.75
13	211.25	215.75	210	214.5	211.25	215.75

get accidentally radiated, minimal disruption to air navigation will occur. Even so, FCC sets radiation leakage compliance standards for theregulated at frequencies used by air navigation equipment. The FCC sets tight frequency standards, so that even if the signals get accidentally radiated, minimal disruption to air navigation will occur. Even so, FCC sets radiation leakage compliance standards for the cable systems to ensure that leakage is low enough not to interfere. Here are the regulations for the frequency measurements of channel carriers:

Regulation: FCC 76.605 (2)

Regulation text: "The aural center frequency of the aural carrier must be 4.5 MHz ± 5 kHz above the frequency of the visual carrier at the output of the modulating or processing equipment of a cable television system, and at the subscriber terminal.

Regulation: FCC 76.612 Cable Television Frequency Separation Standards

Regulation text excerpts: "All cable television systems which operate in the frequency bands 108-137 and 225-400 MHz shall comply with the following frequency separation standards: (a) In the aeronautical radio communication band 118-137, 225-328.6 and 335.4-400 MHz, the frequency of all carrier signals or signal components carried at an average power level equal to or greater than 10^{-4} watts in a 25 kHz bandwidth in any 160 microsecond period must operate at frequencies offset from certain frequencies which may be used by aeronautical radio services operated by Commission licensees or by the United States Government or its Agencies. The aeronautical frequencies from which offsets must be maintained are those frequencies which are within one of the aeronautical bands defined in this subparagraph, and when expressed in MHz and divided by 0.025 yield an integer."

"...offset by 12.5 kHz with a frequency tolerance of ± 5 kHz: or...offset by 25 kHz with a tolerance of ±5 kHz..."

In Other Words

- ◆ The aural carrier must be 4.5 MHz ± 5 kHz from its visual carrier
- ◆ In the navigational bands carrier offsets of 12.5 and 25 kHz have a tolerance of ±5 kHz.

Figure 45 shows the frequency relationships for a single channel's carriers and their tolerances. The FCC regulations require ±5 kHz for the visual carriers within aeronautical navigation bands, but the accuracy is not enforced elsewhere in the cable frequency band. FCC regulations for broadcast transmission of the TV carrier is tighter, at ±1 kHz. The television industry tries to maintain an even more rigorous specification of ±200 Hz. The letter of the law allows the cable system television signals a bit more frequency latitude, but in practice it is well to keep all channels within the FCC cable tolerance because performance variations in set-top converter and television receiver frequency-tuning accuracy may degrade subscriber performance if channels are allowed to drift.

Absolute Frequency Tolerances

Reenforcing the discussion on absolute and relative measurements, note that most of the intrachannel measures are relative frequency measurements. Even the channel definition states that the visual carrier shall be 1.25 MHz above the lower frequency boundary of the channel. But the receivers and tuners look for absolute carrier frequencies, and only then

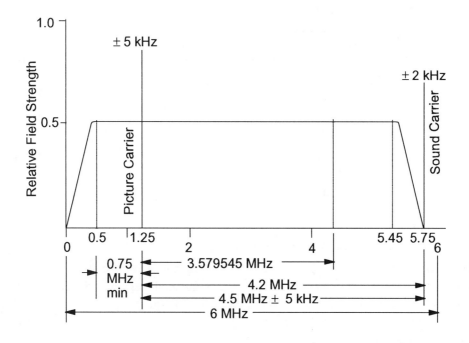

Figure 45. A single channel's frequency parameters and tolerances.

the modulation relative to the visual carrier. Table 4 shows the absolute frequency tolerance in kHz for different frequency translation components of a cable television head end as specified by their manufacturers.

To put these limits into perspective, consider the accuracy that a broadcast transmitter must have according to the CCIR: ±1.0 kHz. One justification for this tight tolerance is the ±10.0 kHz co-channel carrier requirement. Co-channels are carriers broadcast on the same channel allocation within a specified geographic distance. One of these channels is required by the FCC to offset their visual carrier frequency. A ±10 kHz spacing is used to prevent interference with its co-channel. If broadcast tolerances were any greater than ±1 kHz, the establishment and maintenance of co-channels would be tough, if not impossible. Co-channel interference measurements within the cable system is discussed in Chapter 11.

Table 4. Frequency tolerances of various cable television components and signals.

Signal Output of	Tolerance	at 100 MHz
Heterodyne processor	±0.0025%	±2.5 kHz
Demodulator-Modulator	±0.0025%	±2.5 kHz
Broadcast transmitter		±1.0 kHz
CATV relay service	±0.0005%	±0.5 kHz
FM station		±0.2 kHz
Subscriber tap or converter		±5.0 kHz

Spectrum Analyzer Frequency Accuracy

How accurate does the spectrum analyzer have to be to measure the tolerances on head end carrier frequencies? Just as in amplitude accuracy, an analyzer's own accuracy must be figured into the readout it provides. But in the case of frequency the computation job is much simpler because the accuracy is specified by a formula. Table 5 shows typical frequency specifications for a spectrum analyzer.

Frequency Measurements Are Easy

It is easy to measure signal frequency with a spectrum analyzer. Place the signal on a reference point, such as the center frequency graticule, and read the center frequency. Or faster, place a marker on the signal and read its value in the marker readout. Accuracy of the frequency is dependent upon several factors, as the uncertainty equations from our generic analyzer shown in Table 5 under the heading "Frequency Accuracy." Better accuracy is available by turning on the marker count function, a built-in frequency counter. The accuracy is specified by the terms under the "Marker Count Accuracy" heading. The accuracy is improved over the uncounted marker by the removal from the accuracy equation of the uncertainty due to the resolution bandwidth and frequency span.

Frequency Reference Error

Common to both accuracy equations is the term frequency reference error. The frequency reference error describes the analyzer's fundamental frequency stability, that is, the ability of the analyzer's internal reference oscillator to remain unchanged in frequency over a

Table 5. Frequency specifications for a typical RF spectrum analyzer.

Frequency Reference Error
Aging: $\pm\ 1 \times 10^{-7}$/year
Setability: $\pm\ 2.2 \times 10^{-8}$
Temperature stability: $\pm\ 1 \times 10^{-8}$

Frequency Accuracy
Frequency span $\leq$10 MHz $\pm$(frequency readout $\times$ frequency reference error + 3.0 % of span + 20% of RBW + 100 Hz)
Frequency span >10 MHz $\pm$(frequency readout $\times$ frequency reference error + 3.0 % of span + 20% of RBW)

Marker Count Accuracy (S/N $\geq$25 dB, RBW/span $\geq$ 0.01)
Frequency span $\leq$10 MHz $\pm$(marker frequency $\times$ frequency reference error + counter resolution + 100 Hz)
Frequency span >10 MHz $\pm$(marker frequency $\times$ frequency reference error + counter resolution + 1kHz) where counter resolution selectable from 10 Hz to 100 kHz

Frequency Span
Range 0 Hz (zero span), 1 MHz to 1.8 GHz
Accuracy $\pm$2% of span, span $\leq$10 MHz; $\pm$3% of span, span >10 MHz

given time or condition. A long time period, such as a 1-year calibration cycle, is sometimes called an aging-frequency reference error. Short time periods, such as a single day, may also be specified. Frequency stability for spectrum analyzers exposed to a wide temperature range, such as encountered in outdoor cable distribution measurements, is specified as temperature stability. Temperature stability reference error tells how much the analyzer will drift over an analyzer's operating temperature range.

CAUTION

For an analyzer to be accurate enough for carrier frequency compliance tests, its frequency reference aging must be at least 4 parts in 10 $^{-7}$ when used with an internal frequency counter.

The values of frequency reference error are given without units because it a ratio of change for the analyzer's built-in standard oscillator for a period of time or change in temperature, often expressed as parts per million, e.g., 3.5×10^{-6}. The error is always multiplied by a frequency term to yield a frequency uncertainty value in ±Hz.

Example 23. Calculating frequency aging.

The aging specification in Table 5 is 1×10^{-7} per year. How much can the analyzer frequency accuracy at 450 MHz change any time during that year?

Simply multiply the aging by the frequency in Hz: $1 \times 10^{-7} \times 1 \times 450 \times 10^{6} = 45$ Hz. The frequency can drift up or down, so this value is written ±45 Hz.

Often a spectrum analyzer option will allow additional accuracy of the frequency reference. In the last example, if the frequency reference error were to be improved by an order of magnitude, from 1×10^{-6} to 1×10^{-7}, the accuracy of the analyzer's frequency readout would be improved proportionally. For the above example, the aging would be ±4.5 Hz, ten times better than the ±45 Hz calculated.

Techniques in Frequency Measuring

Absolute frequency measurements are made two ways with a spectrum analyzer:

- ◆ Use of the trace frequency scaling and/or a marker
- ◆ Use of the analyzer's built-in frequency marker

The marker and frequency counter method is more convenient and faster, but it may take more time. The time is used to open the frequency count gate, compute the answer, and display the results. Since the counter accuracy is dependent upon the precision of the analyzer's reference oscillator, additional expense may be involved in improving its performance.

Here is an example of the first frequency measurement technique, the use of the trace display:

Example 24. Measuring a Translated Visual Carrier.

Channel 44 is a signal taken off-air and then translated to a new channel at the head end with a demodulator-modulator. Calculate the accuracy of the channel 44 visual carrier when measured at the output of the modulator with the analyzer's trace marker.

Channel 44 visual carrier is at 343.25 MHz. With the analyzer's span at 6 MHz and its resolution bandwidth at 300 kHz, place the trace marker at the peak of the visual carrier. From Table 5 the accuracy of this marker number is ±(frequency readout x frequency reference error + 3.0% of span + 20% of RBW + 100 Hz). Filling in the numbers to get the uncertainty gives ±(343.25 MHz × 1 × 10^{-7}/year + 0.03 × 6 MHz + 0.2 × 300 kHz + 100 Hz) = 34.325 Hz + 180 kHz + 60 kHz + 100 Hz ≅ ±240 kHz.

The ±240 kHz uncertainty is insufficient for assuring the frequency output of even the most inaccurate of the head end components.

The sources of uncertainty when using the analyzer's trace and marker readout is almost entirely due to the analyzer's span and resolution bandwidth. If the spectrum analyzer has a built-in frequency counter, these terms will not affect the uncertainty. Here is the same measurement made with a counter.

Example 25. Counting a Visual Carrier.

Measure the channel 44 visual carrier using the analyzer's counter marker and calculate its accuracy.

Channel 44 visual carrier is at 343.25 MHz. With the analyzer's span at 6 MHz and its resolution bandwidth at 300 kHz, place the trace marker at the peak of the visual carrier, and turn on the analyzer's frequency counter. Set the counter resolution to 100 Hz. From Table 5 the accuracy of this counter marker number is ±(marker frequency × frequency reference error + counter resolution + 100 Hz). Filling in the numbers to get the uncertainty gives ±(343.25 MHz × 1 × 10^{-7}/year + 100 Hz + 100 Hz) = 34.325 Hz + 100 Hz + 100 Hz ≅ ±234 Hz.

Figure 46 shows the analyzer marker readout of 343.2621 MHz. The computed accuracy for this reading places the actual channel carrier at 343.2621 MHz ± 234 Hz, or 343.2621 ±0.000234 MHz. In other words, the carrier is actually between 343.2623 MHz and 343.2619 MHz. This accuracy is sufficient to assure that the demodulator-modulator has placed the channel properly within the ±5.0 kHz window of the FCC regulations for aeronautical bands.

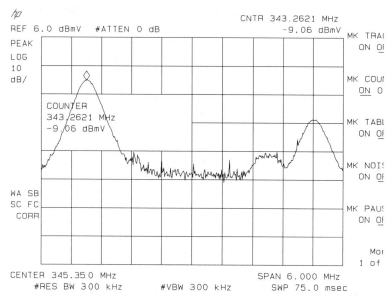

Figure 46. Use of the counter marker to read the visual carrier frequency.

Testing Head End Frequency Converters

In order to test the head end frequency conversion components to their manufacturer's specifications, as in Table 4, the analyzer would have to be about ten times better in its uncertainty. For example, a typical frequency accuracy for the demodulator-modulator is ±0.0025%, which is ±8.58 kHz at the channel 44 visual carrier frequency, 343.25 MHz. To guarantee the demodulator-modulator performance, the analyzer would have to have ±8.58 kHz/10, or ±858 Hz uncertainty. Example 25 demonstrates that, in this case, the analyzer's accuracy is sufficient to test the head end components to their manufacturer's specifications.

To test accurately, the analyzer must be at least ten times better than the compliance specification.

Relative Frequency Measurements

Keeping the Aural Carrier in Its Place

The position of the aural carrier is specified to be 4.5 MHz above the visual carrier, to a tolerance of ±5 kHz. This is a relative frequency measurement. In order to make the measurement to the required tolerance, the analyzer must be at least ten times more accurate than the ±5 kHz specification, that is, an uncertainty better than ±500 Hz. Relative frequency measurement techniques depend upon the performance and features of your spectrum analyzer. Here are the most common techniques for measuring relative frequencies:

◆ Place both signals on the same trace and measure the difference between them with two trace markers.

◆ Measure each signal absolutely, then calculate the difference.

Marker count gives the best accuracy for either of the above techniques, but the following examples will explore the measurements with and without the counter feature.

Example 26. Aural Carrier Frequency on the Same Trace

Use the analyzer markers to read the difference between the visual and aural carriers of a single channel.

Place the channel in a single 6 MHz span and use the 300 kHz resolution and video bandwidths as before. Place a single marker on the visual carrier and a second, or relative marker, on the aural carrier peak. Read the difference in frequency as shown in Figure 47(a).

The reading is 4.485 MHz. This is 15 kHz lower than the 4.5 MHz specification. To determine how much of this 15 kHz is due to analyzer uncertainty, use the analyzer specifications for relative frequency accuracy found in Table 5 under the heading

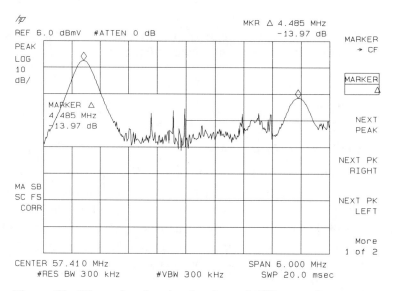

Figure 47a. Measuring the visual and aural difference frequency with plain markers.

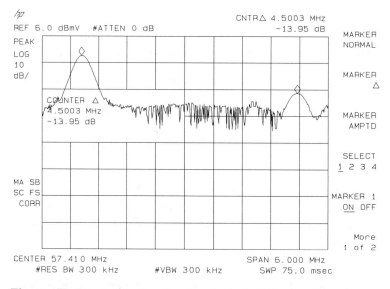

Figure 47b. Measuring the visual and aural difference frequency with counter markers.

"Frequency Span." For spans of less than 10 MHz the accuracy between two points is ±2%. For a separation of 4.5 MHz this is 0.02×4.5 MHz, or 90 kHz! Far too much to measure the aural-visual carrier difference. Narrowing the span to reduce the error won't help, because one of the signals will be off the display.

To improve the accuracy of the markers, use the counter feature. Place a single marker on the carrier with the counter on, then move the second marker to the aural carrier. The reading in Figure 47(b) is 4.5003 MHz, much more believable. The accuracy of this marker-counter relative measurement is the sum of each counter marker's absolute accuracy. From Example 25 these are less than ±300 Hz each, for a total of ±600 Hz uncertainty. In fact, in the reading of 4.5003 MHz, the 300 Hz is probably analyzer measurement error and not aural carrier misplacement.

Conclusion: the use of markers on a single trace is only accurate enough for compliance frequency measurements when the counter function is used.

The second way to measure differential frequencies is to measure each signal absolutely and take the difference. Although the measurement in the second part of Example 26 appears to be a marker comparison, it is really the difference of two independent readings. The analyzer's counter counts only the signal it is on. When the second marker is moved to the aural carrier, the aural carrier is counted, but the analyzer keeps a running computation of the difference between the first counter measurement, which is stored away in memory, and the second real-time count. The difference is readout on the display, saving you the trouble of writing down two frequencies and subtracting them.

If the analyzer you are using does not have a frequency counter, the only way to measure signal differences to the require accuracy is to measure each signal separately in the narrowest span practical, then subtract the two numbers. Here is an example.

Example 27. Aural Carrier Frequency Without a Counter

Measure the difference between the visual and aural carriers of a single channel with an analyzer without a frequency counter.

Span down to about 50 kHz around the visual carrier. Remember to place the resolution and video bandwidths in their automatically coupled modes to allow the

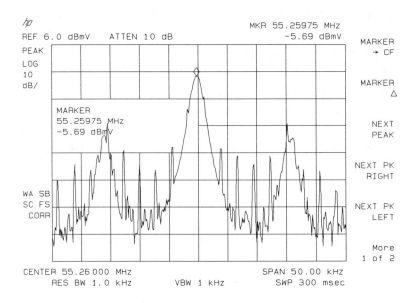

Figure 48a. Measurement of the visual carrier without a counter.

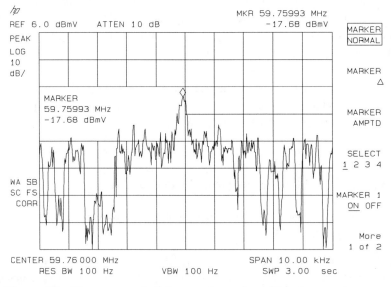

Figure 48b. Measurement of the aural carrier without a counter.

analyzer to resolve signals in this narrow frequency span. (Often this is called auto-coupled bandwidths.) The resolution bandwidth will be 1 kHz. Figure 48.

The visual carrier in (a) is easily measured because enough carrier energy is present as a CW signal. Place a marker on the carrier peak to read its frequency. The reading is 55.25990 MHz. The aural carrier in (b) is a bit more tricky, since you must capture the aural carrier at rest, that is, when there is little or no modulation. Place the carrier roughly in the center of the display, with the span set to 10 kHz. Set the analyzer sweep trigger to single to freeze each sweep. Trigger a single sweep until a suitable tracing of the FM carrier comes along. Use the marker to read its frequency, in this case 59.76013 MHz.

Calculate the difference frequency by subtracting the readings: 59.75993 MHz − 55.25990 MHz = 4.50003 MHz, an error of 30 Hz. The excellent accuracy is due to the narrow spans in which the measurements were made. Remember from the absolute accuracy equations in Table 5, under the heading "Frequency Accuracy," that the uncertainty of each absolute frequency reading is proportional to the frequency span. The span accuracy is ±3% for this analyzer, so the frequency accuracy is 0.03 × 10 kHz = ±0.3 kHz. The total measurement uncertainty worst case is twice this value, or ±0.6 kHz, very close to the ±500 Hz tolerance allowed for compliance measurements.

Making relative frequency measurements without a counter can be as accurate as those with a counter if the analyzer and signals under test are stable enough to allow the use of very narrow spans and resolution bandwidths as in the previous example. But the procedure is tedious and time-consuming.

Counter-Measure Limitations

Tips for Using the Frequency Counter

As easy and accurate as the spectrum analyzer's frequency counter is, there are conditions when signals cannot be counted accurately. These are:

- When the signal has too much noise.
- When the signal is not stable.

The frequency counter in the spectrum analyzer must have a signal that is clear and stable enough to lock onto during the counter's gate time. If the signal is too close to the noise displayed by the analyzer, random noise in the analyzer's IF circuits will disrupt the counting process, resulting in a false count and inaccurate frequency readout. One symptom is a frequency count that changes from sweep to sweep. However, false counts may appear stable even though the signal is too noisy. Keep in mind that the analyzer has no way of determining whether the signal being counted is strong enough, that is, high enough out of the viewed noise. You must be aware of the analyzer's signal-to-noise ratio specification for its counter accuracy. This value can be found in the spectrum analyzer's performance specifications under marker counter accuracy. For instance, the off-air channel 7 visual carrier as seen in Figure 49 appears only about 12 dB out of the displayed noise, and yet the measurement of its frequency appears to be correct. The analyzer specifications for frequency-count accuracy, seen in Table 5 under the heading "Marker Count Accuracy," are only valid for signals that are 25 dB or more out of the noise. In this case, the counted signal is accurate, at 175.2401 MHz, even though the signal is closer to the noise than recommended.[1]

TIP	**Even though it is not necessary to make a counter-frequency measurement at the peak of a signal, the counter marker must be well above the displayed noise.**

Why can the analyzer make such an accurate measurement when the signal is so close to the noise? In general, signals such as visual carriers, which have an uninterrupted signal at their core no matter what other modulation is present, are easy to count close to the noise because the analyzer's counter will gate on the largest or most steady frequency vector in the composite signal. The noise obscures the signal's modulation sidebands more than the central CW carrier. Much more will be said about measuring low-level signals in the chapter on coherent disturbances.

If a signal is not stable, the frequency count will be different each time the signal is counted. A signal, such as the television aural carrier, which is frequency modulated, is at its specified frequency only when there is no audio. In other words, it has no continuous carrier to count on, so to speak. However, the FM signal deviates only 25 kHz from its specified center frequency. When we measured the aural carrier with the counter in Example 26, the analyzer's resolution bandwidth was set to 300 kHz, wide enough to contain all the aural carrier sidebands whether modulated or not. Since the aural carrier

---

[1] Channel 7 is a negative co-channel in its area. Therefore, the visual carrier is lower than the FCC-specified 175.25 MHz by 10 kHz, or the 175.24 MHz shown in the figure.

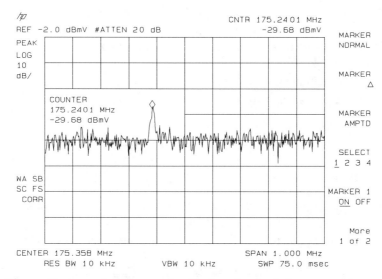

Figure 49. A carrier close to the displayed noise being counted.

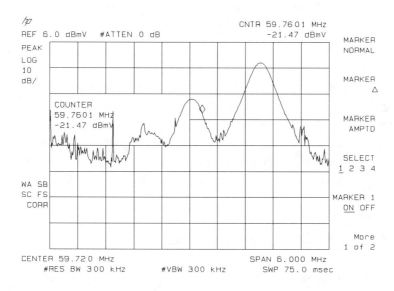

Figure 50. Increasing the counter resolution when measuring FM aural carrier frequency.

spends as much time above the center frequency as below it, the average over the count period will likely be a very good representation of the aural carrier frequency. To improve the chances of an accurate count, make sure the counter marker is in the center of the signal, usually the displayed peak. However, since the aural carrier does not have a continuous carrier component, it is more likely to give false counts if the signal-to-noise ratio is below the analyzer's specification marker counter specification.

Example 28. Measure the aural carrier frequency.

In a 6 MHz span, place a single marker on the peak of the aural carrier and turn on the counter. The readout jumps up and down in frequency only about 100 to 200 Hz even when the marker is placed off to the side of the signal as shown in Figure 50.

TIP	**If the count reading is not changing from count to count, it is probably a reliable reading.**

The stability of the counted signal in Example 28 is a good indication of the counter accuracy. Generally, if the counter is stable, that is, the readout is not changing from count to count, the readout up to the least significant digit can be trusted. For example, if the last digit in a readout of 59.7601 MHz changes from sweep to sweep, you can have confidence that the measurement is accurate to one count of the next highest digit, that is, ±0.001 MHz, or ±1 kHz. If the 0 in the 59.7601 MHz reading were changing, the accuracy would be no better than ±10 kHz.

Summary of Carrier Frequency Measurements

Frequencies of systems need to be measured because demodulators-modulators and processors shift the frequencies of cable signals. The spectrum analyzer can make compliance carrier-frequency measurements. The absolute frequencies of channels need be measured to assure the proper placement of the remodulated channels. The relative frequencies of the aural carriers need to measured to assure proper TV receiver reception and to prevent interference with adjacent channel video. To make frequency measurements accurately with a spectrum analyzer:

- Run the analyzer's calibration routines periodically, or when the ambient temperature changes more than 5°C.
- The analyzer's markers and internal frequency counter simplify measurement procedures.
- Compliance measurements require a tolerance of ±5 kHz.
- Compute the accuracy of your analyzer for absolute frequency measurements using the data sheet equations.
- For consistent compliance measurements, the analyzer must be at least ten times more accurate than the measurement tolerance, or ±500 Hz.
- Relative measurement accuracy when both signals are on the display is dependent upon the analyzer's frequency span accuracy, usually ±3% to ±5%.
- Accuracy of absolute frequency measurements depend upon the analyzer's internal frequency reference specification, usually between 1×10^{-6} to 1×10^{-8}
- Use the analyzer's frequency counter on signals that are stable and free from noise.

Selected Bibliography

- Benson, K. Blair, and Whitaker, Jerry. *Television Engineering Handbook.* Rev. ed., McGraw-Hill, Inc., 1992.
- Deschler, Kenneth T. *Cable Television Technology.* McGraw-Hill Book Company, New York, 1986.
- Engelson, Morris. *Modern Spectrum Analyzer Theory and Applications.* Airtech House, Inc., Portland, OR, 1984.
- Inglis, Andrew F. *Video Engineering.* McGraw-Hill, Inc., New York, 1993.
- Peterson, Blake. *Spectrum Analysis Basics.* Hewlett-Packard Company, Application Note AN 150, Literature No. 5952-0292, Santa Rosa, CA, 1989.

7

Conquering Carrier-to-Noise

In this chapter you'll learn about:

- The nature of noise
- Noise as a signal
- Compliance regulations
- Noise corrections
- Accurate C/N test methodology

Overview

Carrier-to-noise ratio, abbreviated C/N, has been a critical cable television measurement from the beginning. The spectrum analyzer was and is used because of its ease-of-use and convenience; however, the adaptation of the analyzer to noise measurements can seem confusing and laborious. The objective of this chapter is to take the mystery out of spectrum analyzer C/N measurements, and replace the confusion with an orderly analysis of the procedures and a "feel" for the measurement. The goal is to make you feel comfortable and confident when you make carrier-to-noise measurements.

The Nature of Noise

The organized flow of the electrons of an RF signal is disrupted by the disorganized energy of thermal noise, which is caused by heat's action on resistive elements. A television signal is transmitted in and processed by elements that also contribute considerable noise. If the signal becomes weak or the noise increases, the reception loses clarity. As the signal modulation of a television visual carrier sinks into the noise, snow appears over the whole picture. This is illustrated in Figure 51

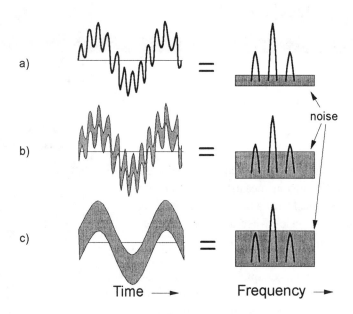

Figure 51. The effect of noise on signal modulation.

On the left are the time domain wave forms of a signal with sinusoidal amplitude modulation. On the right, the spectrum analyzer's view of each signal's frequency domain presentation. The effect is as if the noise were a rising sea engulfing the signal and its sidebands. The sequence from (a) to (c) shows the signal stripped of its sidebands. The sidebands represent the intelligence being transmitted by the central carrier, just as the television signal sidebands carry picture and sound information. As the sidebands are obliterated by noise, so goes the picture.

The Measurement in Brief

1. Connect the analyzer to the system in a span of 2 MHz tune to a place with only noise at the display center, or turn off the carrier modulation.
2. Set the resolution bandwidth to 30 kHz and video bandwidth to 100 Hz.
3. Disconnect the cable input from the analyzer or the preamplifier, if used, and note the noise-level drop.
4. If the drop was between 1 and 10 dB, go to Step 7. If <1 dB, continue.
5. Add a preamplifier with a gain between 20 and 30 dB, and a noise figure ≤ 10 dB.
6. Go back to Step 2.
7. Connect the system input and record the noise level in dBmV from the marker readout.
8. Correct the noise reading for 4 MHz, resolution bandwidth, detection, and equivalent noise power bandwidth by adding 23.24 dB.
9. If the system noise was within 10 dB of the analyzer noise, correct the reading for noise-near-noise.
10. Reinsert the carrier and measure its level in dBmV using resolution and video bandwidths of 300 kHz.
11. C/N is the difference between the results of Steps 8 and 10, in dB.
12. Correct C/N for preamplifier noise figure, if used.

Picture Critique Subjective, But Rules Are Objective

From your subscribers' point of view, noise added to the television picture causes visible fading of the picture. This effect is shown in Figures 52, 53, and 54, using a simulated television picture. The picture on a television receiver with a poor C/N signal is awash in random spots that flicker uniformly about the screen at the television's scan rate. At a C/N near 36 dB the picture loses all detail. This effect is often called snow because of the similarity to a scene obliterated by snowfall.

Subscriber reaction to this noise depends upon how important the picture is to him or her, the type of picture being transmitted, the quality and condition of the TV receiver and/or set-top converter, and the C/N of the delivered signal. Not much can be done about the subjective aspects of picture critique nor the TV receiver, but the noise levels themselves are subject to more exact, that is, objective, C/N testing.

Figure 52. Simulated television picture with >50 dB C/N.

Figure 53. Effect of decreasing the C/N to 43 dB on the simulated television picture.

Figure 54. C/N at 36 dB.

> **TIP**
>
> **Complaints about washed-out pictures can be caused by poor system C/N. But it could be as simple as a poor connection to the subscriber's television or set-top converter!**

FCC Regulations

The FCC regulations governing the specifications for and testing of carrier-to-noise ratio are as follows:

Regulation: FCC 76.605 (a) (7) and FCC 609 (e)

Regulation Text For Technical Standards: "The ratio of RF visual signal level to system noise shall be as follows: (i) From June 30, 1992, to June 30 1993, shall not be less than 36 decibels. (ii) From June 30, 1993, to June 30 1995, shall not be less than 40 decibels. (iii) From June 30, 1995, shall not be less than 43 decibels. (iv) For class I cable television channels, the requirements of paragraphs (a) (7) (i), and (a) (7) (ii) of this section are applicable only to: (A) Each signal which is delivered by a cable television system to subscribers within the predicted Grade B contour for that signal; (B) Each signal which is first picked up within its predicted Grade B contour; (C) Each signal that is first received by the cable television system by direct video feed from a TV broadcast station, a low power TV station, or a TV translator station."

Regulation Text For Measurements: "System noise may be measured using a frequency-selective voltmeter (field strength meter) which has been suitably calibrated to indicate rms noise of average power level and which has a known bandwidth. With the system operation at a normal level and with a properly matched resistive termination substituted for the antenna, noise power indications at the subscriber terminal are taken in successive increments of frequency equal to the bandwidth of the frequency-selective voltmeter, summing the power indications to obtain the total noise power present over a 4 MHz band centered within the cable television channel. If it is established that the noise level is constant within this bandwidth, a single measurement may be taken which is corrected by an appropriate factor representing the ratio of 4 MHz to the noise bandwidth of the frequency-selective voltmeter. If an amplifier is inserted between the frequency-selective voltmeter and the subscriber terminal in order to facilitate this measurement, it should have a bandwidth of at least 4 MHz and appropriate corrections must be made to account for its gain and noise figure. ..."

What This Means, Practically Speaking

Here are the regulations in abbreviated terms..

- The demands of C/N testing are getting stricter, ultimately to a compliance level of 43 dB.
- C/N levels are to be measured at the set-top converter output.
- Very few television channels are exempt from testing.
- Measurements must be made within the channel bandwidth, therefore with the television modulation off.
- Noise levels must be referenced to a 4 MHz bandwidth.

In addition, the FCC Rules say "...measurements made in accordance with the NCTA Recommended Practices for Measurement on Cable Television Systems ... may be employed." The measurement procedures can also be provided by cable television industry experience and mutual agreement.[1]

---

[1] Organizations, such as the National Cable Television Association, in cooperation with the FCC and experienced technical advisors from the cable industry have published procedures that are accepted by the FCC for compliance testing.

Set-Top Converters

The desire to measure the C/N at the output of the subscriber's TV converter has similar drawbacks as in the other tests of this regulation, that is, the impracticality of making the measurements in the subscriber's home. With other tests, such as the visual carrier levels, it is sufficient to use 30 meters of cable from the subscriber's tap to represent the point of delivery. But the TV converter has a significant effect on the final C/N delivered to the subscriber, and therefore cannot be left out of the test procedure. However, the NCTA recommends a procedure that derives a set-top converter correction factor based upon the converter manufacturer's specifications, that, when applied to the tap carrier-to-noise ratio, represents the C/N level at the TV receiver. In this chapter, the measurement of carrier-to-noise with the spectrum analyzer does not include the converter correction factors.

CAUTION — **C/N measurements in this book do not include the television set-top converter.**

Measuring the Carrier

As the term carrier-to-noise ratio implies, C/N is a comparison of the carrier level to the noise of the system. The carrier level is measured using the procedures given in Chapter 4. No surprises here! However, there are a few techniques to reemphasize.

Every dB Counts

A single dB of uncertainty in the measurement of the visual carrier directly affects the C/N level. As regulations get tighter and tighter, the need for consistent and accurate measurements to the sub-dB levels become more important, especially when your system performance is close to the compliance levels. Overload prevention is critical since compression can cause significant amplitude error. To ensure system compliance, guardbands need to be set based upon the accuracy of your test equipment, and level tested with these guardbands in place.

TIP — **Since very dB counts when measuring C/N, follow best accuracy practices when measuring the visual carrier amplitude.**

Use Correct and Consistent Procedures

Another important lesson is to be consistent in the way the carrier is measured, that is, use the same analyzer control settings recommended in Chapter 4 without variation.

The visual carrier needs to be measured with its modulation on, even though the noise measurement, under compliance rules, requires the modulation be turned off for the noise-level reading.

For best accuracy in making relative amplitude measurements, the highest carrier is left at the reference level. The same is true for the measurement of C/N. As you will see, the procedure is to measure the carrier at the reference level, and then without changing the reference level setting, make the noise level measurement. This procedure avoids uncertainties that would make C/N much less dependable. Examples illustrate this process later in this chapter.

In summary, to make the carrier amplitude level measurement for the C/N:

- Use the recommended control settings to make an accurate visual carrier level reading and bring carrier peak to the reference level.
- Measure a modulated carrier.
- Assure that analyzer is operating free from overload.
- Apply guardband to the amplitude to remove as much uncertainty from the analyzer's reading as possible.
- Do not change the reference level when making the noise measurement unless required to raise noise out of lowest display division.

Measuring Noise

C/N is the ratio of carrier level to noise level. The remainder of this chapter discusses how the analyzer's operation and features are put to work to make accurate and consistent noise-level measurements even though the spectrum analyzer's primary purpose has been to make CW signal level and frequency measurements. Once the noise level is known, carrier-to-noise ratio is simply the ratio of the peak of the carrier to the average of the system noise, a simple subtraction of the power dB values.

Before using the analyzer to measure noise, let's look at the nature of noise as a signal.

Noise Is Everywhere

What is noise in the environment of electrical signals? Noise is a signal, with power and frequency characteristics, just like any other signal. In other words, it has a level and frequency response, and it can be amplified, transmitted, and measured. All electrical components generate noise. Fundamentally, the electrical resistance of a component always generates noise power if the component's temperature is above absolute zero degrees Kelvin. Theoretically, at zero degrees Kelvin all molecular activity stops, and no noise power is produced. At room temperature noise is generated by resistance across the component leads as a random voltage. This voltage can be measured, but only by special devices much more sensitive than a spectrum analyzer or signal level meter.

Passive components, such as resistors, capacitors and inductors, generate noise according to their temperature, illustrated in Figure 55.

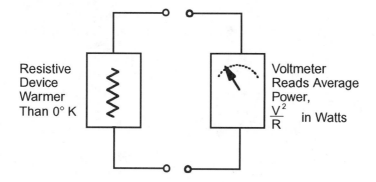

Figure 55. Noise generated by a component.

The voltage at the component leads across the component's internal resistance generates power. Active devices, such as transistors, add noise to this thermal noise. Thus, when any noise is amplified, the noise output is increased by more than the gain of the amplifier. This means that when noise is measured through a chain of active devices, the measured level must be corrected for noise contributions from the amplifiers used to make the measurement. Such is the case when a preamplifier is used to increase the spectrum analyzer's noise measurement sensitivity. It is not as complicated as it sounds, as the following sections will show.

Noise Is Measured as Power over a Frequency

To help understand how the spectrum analyzer measures noise, look at the nature of the noise signal. The randomness of noise gives it a theoretically infinite frequency spectrum. At any one point in the time or frequency domains, truly random noise appears as a continuous range of amplitudes rather than a single value. This is illustrated in Figure 56.

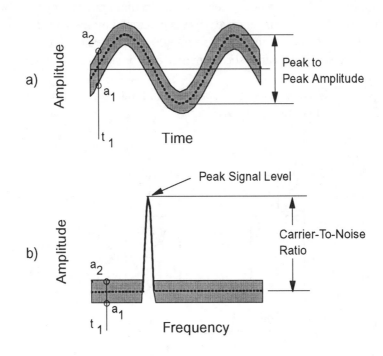

Figure 56. Signal and noise in the (a) time and (b) frequency domains.

The signal and noise are shown in the time and frequency domains. In (a) a sinusoidal signal in the time domain has its waveform is thickened by the random amplitude variations of noise. The dashed line represents the average of the noise signal, that is, the average of the sine wave itself. At time, t_1, the signal amplitude may be anywhere from a_1 to a_2. In (b), the frequency domain of the same RF signal, represents a CW signal whose amplitude represents the RMS signal power. The dashed line is the average of the noise as in (a). At frequency, f_1, the signal amplitude may be anywhere from a_1 to a_2.

Noise is always measured in terms of a frequency bandwidth, that is power per Hz.

The power of a noise signal is quantified as an average amplitude over a specified frequency range. The frequency range is called its IF noise equivalent bandwidth. The power level is called noise power density, and is specified as dBmV per a IF noise equivalent bandwidth in Hz. Because the television receiver has a bandwidth of about 4.2 MHz, C/N compliance tests require the measured noise power to be referenced to a 4 MHz IF noise equivalent bandwidth. The units of noise power density are dBmV/4 MHz.

Measuring Noise with a Spectrum Analyzer

A Spectrum Analyzer Is Calibrated to Measure CW Signals

But can be adapted to measure many other types of signals, including noise. The spectrum analyzer is calibrated to measure the RMS power of a CW signal. The CW signal is presented on the display in the shape of the analyzer's resolution bandwidth filter. This filter, located in the analyzer's primary signal processing circuits, the IF, allows more than one frequency component to pass to the detector at a time.

The analyzer measures the noise that passes through the resolution bandwidth filter, so if the filter is changed, the noise level changes too.

This attribute of the analyzer's resolution filter makes it valuable in making noise power measurements as Figure 57 illustrates. The noise input to the analyzer in (a) is filtered by the IF resolution bandwidth in (b). The level of the remaining IF noise equivalent in (c) is determined by the input noise level and the shape of the IF resolution bandwidth filter. This noise power is detected, filtered by the video filter, and displayed as a noise level on the spectrum analyzer.

Smoothing the Random Amplitude Variations of Noise

The spectrum analyzer displays this noise as a ragged line across the lower portion of the display as shown, along with a CW signal, in Figure 58. The way an analyzer displays noise is explained in more detail in Appendix D.

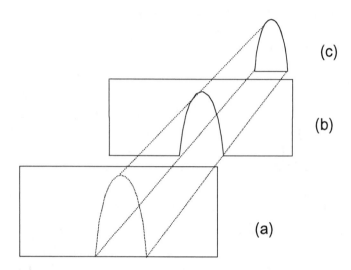

Figure 57. Noise as it appears in the spectrum analyzer IF.

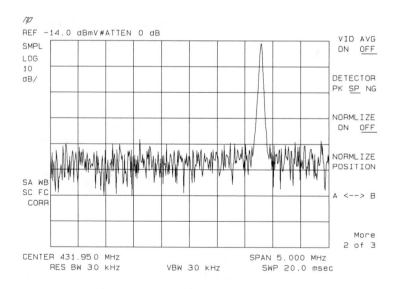

Figure 58. Noise and CW signal displayed together.

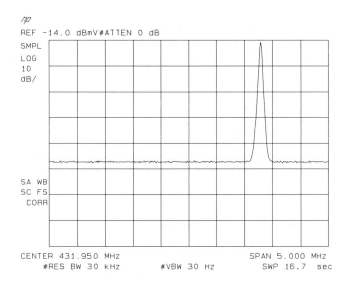

REF -14.0 dBmV#ATTEN 0 dB

SMPL
LOG
10
dB/

SA WB
SC FS
CORR

CENTER 431.950 MHz   SPAN 5.000 MHz
       #RES BW 30 kHz   #VBW 30 Hz   SWP 16.7 sec

Figure 59. Video filter effect on noise.

TIP

The video filter, like a low pass filter, averages the noise, making it easier to read.

Since adjacent frequency points are averaged to similar levels, the effect is to produce a single line across the analyzer's display that represents noise power density over frequency. Figure 59 shows the effect of a lower video filtering setting on the noise displayed in Figure 58.

Noise Power Density

Converting the Analyzer Reading to a 4 MHz Bandwidth

The noise power density is the noise power level read on the spectrum analyzer, corrected by the resolution bandwidth ratio. The FCC regulations require the noise power density referenced to 4 MHz. For an analyzer resolution bandwidth of 30 kHz, the correction is 21.25 dB. Add the correction to the analyzer reading of noise to get the noise power density value relative to 4 MHz.

In C/N, the noise must be measured relative to a 4 MHz bandwidth. The correction for a 30 kHz resolution bandwidth is to ADD 21.25 dB to the noise reading.

Sampling Detection Used for Noise

The spectrum analyzer is set up to make CW signal measurements using a peak detector. Noise, being a random variation of voltage at each frequency, requires the use of the analyzer's sample detector.

Use sample detection for measuring noise, and peak detection for measuring carrier levels.

Automatic Noise Measurement

Most spectrum analyzers are equipped with a noise measurement feature called the noise marker, or noise-level marker. This is a special marker function that sets the analyzer to a sample detection mode, makes additional corrections for log and detection errors, and reads out the computed IF noise equivalent density for a 1 Hz bandwidth. This is a convenient tool for many noise measurements, but not all the corrections necessary for cable television noise measurements are made with this general-purpose function. And, since the corrections and computations are hidden away in the analyzer's memory and computer, there is more to learn about how the analyzer measures cable system noise by avoiding its use for now.

The corrections made by a built-in analyzer noise marker can result in unreliable and inaccurate C/N. If you are unsure about its definition, don't use it.

Summary of Spectrum Analyzer Noise Measurement Practices

If you are concerned that you have missed anything, don't worry; no attempt has been made to make cable system noise measurements yet. The next section satisfies that need. The information thus far serves to introduce you to spectrum analyzer tools that enable accurate noise and carrier-to-noise measurements to be made. Here is a summary of the guidelines for making noise measurements with the spectrum analyzer.

- Noise power density is always stated with a frequency bandwidth.
- C/N is calculated for a noise-power bandwidth of 4 MHz.
- Noise power density is computed using the analyzer's resolution bandwidth.
- Video bandwidth helps smooth the noise for consistent readings.
- Sample detection is required for accurate noise averages.
- Analyzer's built-in noise marker makes automatic corrections, but not all the ones necessary.

TIP	**Additional information on these topics is in Appendix D.**

Is the Noise from the System?

The C/N measurement requires that the system noise be measured. Up to now the source of measured noise has not been identified. How can you tell if the noise displayed is system noise? Simple–just watch the level as the input to the analyzer is disconnected. In Figure 60 the analyzer has been set to read noise over a 2 MHz span in a portion of the cable system that has no channels. Half way through the display the input is disconnected. The drop in noise level demonstrates that the displayed noise on the left is due to the system. The noise remaining is the noise generated within the analyzer itself. The following example demonstrates how to get the display shown in Figure 60.

TIP	**Not all the noise you see comes from the input.**

Example 29. Is that system noise?

Determine whether the noise observed on the spectrum analyzer when connected to the cable system is system or analyzer noise using the disconnect test.

Tune the spectrum analyzer to a place within you system spectrum where there are no channels, but below the highest channel. Span down to about 2 MHz or less. With a 2 MHz span the resolution bandwidth is set at 10 kHz. The noise displayed has so much amplitude variation that some video filtering is necessary. Set the video filter to 30 Hz or less. The sweep time begins to slow down as you narrow the video filter. This is done automatically by the analyzer to keep the CW signal calibrated. It also serves to allow time for sufficient digital processing of the noise points at each frequency across the span, so leave the slow sweep speed. In Figure 60, this is 20 seconds full span. The noise trace is not very smooth, but it is sufficiently flat to test for the presence of system noise.

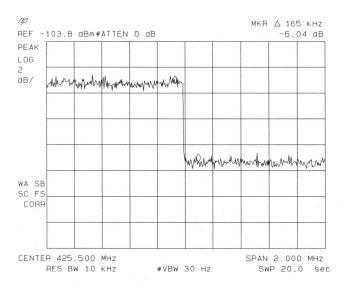

Figure 60. Determining if the noise is from the analyzer's input with the disconnect test.

Next, change the amplitude scale for better resolution of the changes in noise in order to see the noise change of 1 or 2 dB resolution. With a 10 dB per division scale it may not always be possible to tell how far the noise changes. You need to see at least 3 or 4 dB of change; otherwise the change could be attributed to the noise itself. Move the signal up higher on the display using the reference level control. As the amplitude scale is changed, the scale changes with the reference level as a fixed point. So if the trace is not brought near the top, it disappears off the bottom when changing the amplitude scale. Change the amplitude scale to 2 dB per division.

Single sweep the analyzer and disconnect the input while the sweep is going. The single sweep serves to freeze the display so that markers can be put on each side of the noise drop. In this figure the drop is 6 dB.

Disconnect Test for System Noise

This determination of noise drop, called the disconnect test, can be somewhat sloppy in technique because it serves only as a first look to see if the analyzer is capable of making the measurement at all. With a drop of 1 dB or less, it is necessary to boost the analyzer's noise measurement capability. If the noise drop is greater than 1 dB but less than 10 dB, the drop has to be measured very carefully to make noise-power corrections. If the noise drop is more than 10 dB, you can continue with the measurement of noise without concern for the analyzer's noise power measurement accuracy. What was that again? If the disconnect test noise drop is

* > 10 dB, the analyzer can make noise power measurements as is, and without correction factors for internal noise.
* < 10 dB, but > 1 dB, the analyzer is capable of noise measurements with amplitude corrections. The noise difference must be measured accurately.
* < 1 dB, the analyzer noise measurement capability must be improved.

TIP   **The disconnect test determines whether you have to add a preamplifier and/or corrections to your noise measurement procedure.**

Noise-near-noise is the term used when corrections for the analyzer's internal noise are required, as is the case for the second and third conditions above. Let's take each of the these cases and their solutions in order.

When System Noise Is >10 dB Higher Than Analyzer Noise

The only caveat is determining that the system noise is more than 10 dB higher than the analyzer noise. If the signal is close to the 10 dB differential, play it safe and follow the procedures in the next section. Then, if the difference turns out to be >10 dB, no error correction is necessary.

When System Noise Is <10 dB, but >1 dB Higher Than Analyzer Noise

Noise-Near-Noise

This is the case where the noise of the analyzer and system blend to give a composite value, a so-called noise-near-noise condition. The noise contributes according to each level. Since the system noise is the value sought, the analyzer contribution to the composite has to be determined by calculation and verified by separate calibration of the spectrum analyzer. The higher the system noise compared to the internal analyzer noise, the less the analyzer noise influences the displayed composite value. For higher system noise levels, where the difference between the connected and disconnected analyzer noise reading is close to 10 dB, the correction is small. For low disconnect test changes, the correction is high. This is illustrated in Figure 61.

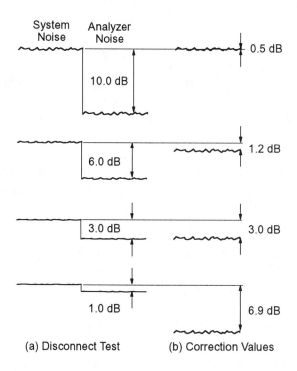

Figure 61. Disconnect test results (a) for different levels of system noise. The actual system noise level for each is shown in (b) as if the level could be displayed on the analyzer.

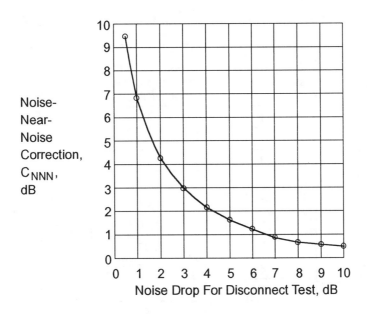

Figure 62. Correction for noise-near-noise using the disconnect test results.

| TIP | **Over 10 dB disconnect test requires neither a preamplifier nor noise-near-noise correction.** |

In the left column of Figure 61 are shown four levels of system noise changes as displayed on the spectrum analyzer. On the right are shown the levels of the actual system noise, as if they could be displayed on the analyzer. The actual levels are calculated values based upon how the analyzer displays noise added to its internal noise. Accuracy is discussed at the end of this chapter.

The exact correction values are shown as a graph in Figure 62 and Table 6. Correction values are plotted for disconnect test noise changes along the horizontal axis. The disconnect test must be done with as much precision as possible because all errors are translated directly to the final C/N results. In addition, small noise differences as shown on left of Figure 62 cause high changes in the error correction, requiring extra care in disconnect test procedures with small differences. Example 30 will help to explain the procedure and the use of the noise-near-noise graph.

Table 6. Value that make up the curve in Figure 62.

Noise drop when disconnecting signal, dB	Amount to be subtracted from measured cable noise level, dB
1	6.87
2	4.33
3	3.02
4	2.20
5	1.65
6	1.26
7	0.97
8	0.75
9	0.58
10	0.46

Example 30. Correct for a noise-near-noise level reading.

Determine the actual system noise by performing the disconnect test.

Follow the instructions in Example 29 to get a display of the disconnect noise drop, but use a resolution bandwidth of 10 kHz and a video bandwidth of 10 Hz. Also set the analyzer's detector for sample mode in order to get a more accurate average of the noise power. Figure 154 in Appendix D shows the results. Note the smaller noise swings of the noise both with the system connected, on the left, and with the system disconnected. The left side of the display also has some composite distortion signals. Place a single marker on a point on the left noise plateau, and move it to a point that represents the average of the noise swings. Avoid using the amplitudes of the composite distortion signals in your eyeball average. Record this value. It is −60.3 dBmV.

Now enable the second, or delta, marker and place it on the analyzer noise level on the right of the cliff, by selecting a point that represents the average. The difference in noise levels is 5.94 dB. On the horizontal axis of the graph in Figure 62 find 6 dB, and read the correction value from the vertical axis as 1.3 dB. Tenths of a dB are close enough for the accuracy of this correction.

The system noise is the first marker reading, −60.3 dBmV minus the correction, or −60.3 dBmV − 1.3 dB = −61.6 dBmV per 10 kHz. The correction for the 4 MHz bandwidth, from Table 11 in Appendix D, is +26.02 dB. The system noise, corrected for bandwidth and noise-near-noise, is −61.6 dBmV + 26.01 dB = −35.6 dBmV/4 MHz.

TIP

The correction for noise-near-noise always decreases the power of the noise level because the analyzer measures the input noise PLUS the analyzer noise.

When System Noise Is Less Than 1 dB from the Analyzer Noise

Improving the Analyzer's Noise Measurement Capability

Noise in the spectrum analyzer can be high enough to mask system noise in many instances. The most important and least obvious way to improve the analyzer's noise measurement capability is to keep its input attenuator at as low a value as possible. The input attenuator puts impedance directly in the input path, reducing noise power dB for dB of attenuation. The effect on the noise-near-noise disconnect test is dramatic. In Figure 63, the top display is the disconnect test with 10 dB of input attenuation. The lower display, with input attenuator set to 0 dB, shows a significant improvement in the analyzer's ability to display the system noise. If reducing the input attenuation cannot improve the analyzer's noise measurement capability, then external noise amplification must be used.

TIP

When the disconnect test give a drop between <1 dB, you must add a preamplifier of ≤10 dB noise figure and between 20 dB and 30 dB gain.

CAUTION

A 1 to 3 dB disconnect drop yields very poor accuracy. It is recommended that an amplifier be used for disconnects of < 3 dB.

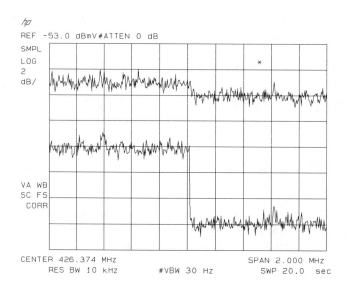

Figure 63. Disconnect test for noise conducted with input attenuation of 10 dB (top trace) and 0 dB (bottom trace).

Using a Preamplifier to Improve the Analyzer's C/N Capability

The spectrum analyzer has a certain level of noise generated with in it. This has been evident in the previous examples, where noise from the input is compared to the internal noise with the disconnect test. Boosting the noise signal to the input of the analyzer allows it to read noise well above its own noise level.

Preamplifier's Effect on Noise

A preamplifier is a good candidate for such a task, but it is not quite that simple. All active devices add noise to the signals passing through them. When a signal and noise are amplified, the C/N through the preamplifier degrades. Gain is the amount of power gained by the signal and the noise as they are amplified. The noise figure is a measure of the preamplifier's degradation of the carrier-to-noise. This can be seen graphically in Figure 64. Signals (a) and (b) represent the spectrum analyzer's view of the input and output of a carrier, respectively. Each signal is accompanied by system noise. The carrier-to-noise ratios of input and output are C/N_{IN} and C/N_{OUT}. The preamplifier applies its gain to the carrier signal, as shown by the relative amplitudes of (a) and (b). The input noise is amplified by the preamplifier's gain plus the noise figure. This is represented by the differences between the signals' absolute noise powers. Since the noise power increases more than the carrier level, C/N_{OUT} is lower than C/N_{IN}.

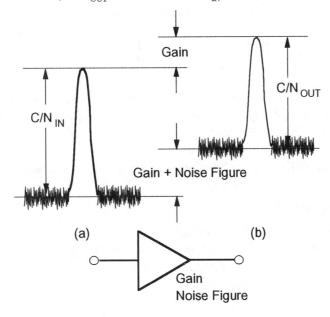

Figure 64. How gain and noise figure affect the carrier-to-noise through a preamplifier.

The job of the preamplifier is to boost the signal and system noise up to where the spectrum analyzer can get a good look, that is, where the analyzer sees noise of the system 1 dB or more higher than its own noise. If the preamplifier has too high a noise figure, the spectrum analyzer measures the preamplifier noise rather than the system noise, masking the system noise. If the gain is too high, the analyzer can be overloaded, and compression renders the C/N inaccurate. A balance of preamplifier noise figure and gain is required. That is why it is recommended that preamplifiers have a 10 dB maximum noise figure and 20 to 30 dB gain.

The preamplifier must not distort the signal or noise levels. This happens when the amplifier is operated in its nonlinear region similar to the discussion in Chapter 4 on internal spectrum analyzer distortion caused by overloading its input mixer. Figure 24 in Chapter 4 also describes overload effect with a preamplifier. Fortunately, overload in the preamplifier is as easy to test as it is for the analyzer. The section on overload below shows how to test both analyzer and preamplifier at the same time.

Using the above guidelines for preamplifier gain and noise figure does not guarantee that the analyzer can make the C/N measurement. As stated, the disconnect test must be passed, this time with the input to the preamplifier being disconnected. When the input is disconnected from the preamplifier, the analyzer displayed noise level drops more than 1 dB, the preamplifier is doing its job, just as in the Example 29. But the analyzer measures the carrier-to-noise as changed by the preamplifier.

The C/N ratio at the preamplifier's output needs to be determined before the preamplifier's noise contribution can be computed. Its error contribution is discussed in this chapter's section on correction factors.

But a proper preamplifier needs to be selected. What are the guidelines?

Selecting a Preamplifier

Many commercial preamplifiers are available for C/N testing. Some spectrum analyzer manufacturers provide external preamplifiers as accessories, or they are built into the spectrum analyzer itself for testing accuracy and convenience. Here is a summary of the desired preamplifier characteristics:

- Cover the entire frequency range of your system, that is, be broadband.
- Operate in its linear region, that is, not overloaded.
- Gain between 20 and 30 dB.
- 75 Ω input and output impedance.
- Noise figure <10 dB, with the value's uncertainty specified.

The last point is important. Since the preamplifier noise figure computes the system C/N, any uncertainty it has is imposed directly on the final C/N result. More on uncertainties later.

Measuring the Preamplifier Gain and Noise Figure

The gain of the preamplifier can be measured easily with the spectrum analyzer. The noise figure is possible to measure, but beyond the scope of this text. You must trust the manufacturer to provide accurate noise figure values with the product. Often these parameters are written on the individual preamplifiers, indicating that they have been tested and documented during the manufacturer's final test. The preamplifier document-ation should show the accuracy for gain and noise figure in terms of a ± dB value.

TIP

Get the preamplifier noise figure from the manufacturer, but test the gain yourself.

Here is how to measure the gain of the preamplifier.

Example 31. Measure preamplifier gain.

To measure the preamplifier gain simply insert the preamplifier between a known signal and the spectrum analyzer, and measure the signal increase in dB. Place a high visual carrier in the center of the 6 MHz display, and set the resolution and video bandwidths to 300 kHz. Bring the signal near the top of the display with the reference level control, keeping the attenuator at the value set when checking overload. If a tunable bandpass filter is used, tune the filter to maximize the carrier signal and readjust the signal amplitude on the analyzer.

Use the marker peak and marker to reference level functions to set the signal maximum on the reference level. If your analyzer does not have these functions, set the amplitude scale to 2 dB per division, and use the reference level control to bring the signal up to the top graticule. Record the signal level.

Remove the preamplifier from the circuit, and adjust the reference level to bring the signal to the reference level again. Record the level and subtract from the amplified level to get the gain of the preamplifier at this frequency.

It is important to remove the preamplifier completely from the circuit, rather than just to turn off its power supply. A preamplifier without power attenuates the signal through it, giving a false reading of the carrier.

It is good practice to evaluate the preamplifier gain at three frequencies across the system to measure its frequency response. If wide variations of gain are seen, more carriers should be measured to confirm the preamplifier's frequency response and gain. If the gain is out of specification, the preamplifier may be damaged. Have it tested to factory specifications.

Guarding Against Overload

The biggest danger when adding a wide bandwidth preamplifier to the input of a spectrum analyzer is that the analyzer can be overloaded. As discussed in Chapter 4, compression of the visual carrier causes it to be read lower than actual. In C/N measurements this leads to a C/N that is worse than expected. This is compounded by the tendency of an analyzer operating in its nonlinear range to generate a great deal of internal distortion, which, by the nature of the cable spectrum, can appear as noise, causing noise to appear higher than actual. High noise means lower C/N. Review Chapter 4 for the procedures of overload testing. If your system is subject to overload, then a tunable bandpass filter must be added to the test set. Figure 65 shows the location of the filter.

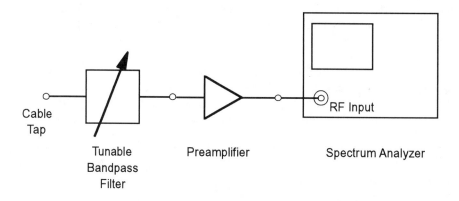

Figure 65. A tunable bandpass filter is required to prevent spectrum analyzer overload with a preamplifier.

TIP

Test for compression overload, and correct with a bandpass or multibandpass filter in front of the preamplifier.

Final Corrections

Just when you thought it was time to start the actual, final measurements, there are two additional corrections that need discussion. First is the correction of the noise level for the actual shape of the IF filter. The second correction is an adjustment for the analyzer's log amplifier and detection.

Correcting for Resolution Bandwidth Shape

The spectrum analyzer resolution bandwidth filters are designed for the accurate and speedy display of CW signals. Noise-density levels depend upon knowing the exact value of the noise passed through the filter. Different analyzers have different shaped filters, so that a filter whose 3 dB bandwidth is 30 kHz in one analyzer may present different levels of noise power to the analyzer's detector. A filter's power-density measurement capability is determined by the area under the filter. Figure 66 (a) shows a common filter shape with signal power P in dBmV. This shape, called Gaussian, for its resemblance to the statistical curve for random distribution, is the shape of the analyzer's IF filter. The shape is generally the result of several stages of tuned filters.

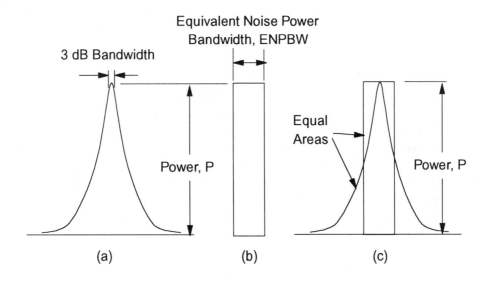

Figure 66. IF noise equivalent power bandwidth filter shapes.

Noise power is flat across the frequency band, so the best way to quantify the noise is to use a flat top, square-sided filter shape as in Figure 66(b). In order for the two filters to pass the same amount of noise, the areas under the shapes must be equal, and their maximum power responses must be at the same level. The level is represented by the value P. The bandwidth of the rectangular filter, called the noise equivalent power bandwidth or NEPBW can be determined if the area is known. This relationship is shown in (c).

TIP

An analyzer with a Gaussian-shaped resolution bandwidth has a filter shape correction of 0.5 dB. Subtract this from the noise power level.

The NEPBW is different from the Gaussian filter's 3 dB bandwidth because the area of each filter shape is distributed differently. Once you know the area under the Gaussian filter shape, the NEPBW is easily calculated. See Appendix D for details on how to measure your analyzer's equivalent bandwidth correction.

Don't confuse this correction with the adjustment of noise power density between the analyzer's bandwidth and the 4 MHz cable system noise bandwidth. The 4 MHz correction is just a mathematical normalization to a standard measurement convention. The NEPBW is an error correction dependent upon the specific analyzer filter characteristics.

Correcting for IF Detection and Logarithmic Conversion

The spectrum analyzer is a voltmeter that displays results in a logarithmic scale. Measured voltage is converted into a logarithmic value, which allows a very wide range of signal and noise powers on the same display. However, measuring noise in this way causes the noise to be read too low by 2.5 dB in most spectrum analyzers. This made up of 1.05 dB from envelope detection of the Gaussian noise distribution, often referred to as the Rayleigh distribution, and 1.45 dB from converting voltages to logarithmic values. Because the effect is due to the log amplifier and the detection, one correction is provided. Computations and test procedures to derive this correction factor are beyond the technical level of this book, but spectrum analyzer manufacturers who serve the cable television industry have fully characterized their products to provide the correction value.

The important lesson about this error correction is to remember which way it corrects the noise density. Since the log detection circuits compress the noise from the top, the noise reading reads lower than actual. So the correction needs to be added to the read value, just like the correction for 4 MHz bandwidth. In fact, the best way to remember the sign on the correction factor is to do the same when converting from an analyzer bandwidth to the 4 MHz bandwidth.

Example 32. Correction for log detection.

Correct the noise power density of the last example for log detection.

The noise power reading is −63.09 dBmV/30 kHz. To correct for the log detection ADD 2.5 dB. The sign is the same as for the 4 MHz power density correction. The value corrected for both log detection and 4 MHz is −63.09 dBmV/30 kHz + 2.5 dB + 21.25 dB = −39.34 dBmV/4 MHz.

This looks like an overwhelming number of corrections and adjustments to make for one noise measurement. And just when you thought it was safe to start measuring, one more correction appears. If a preamplifier was required to make the spectrum analyzer view system noise, that is, pass the disconnect test, then read the following section.

Correcting C/N for Preamplifier Noise Figure

The logic behind the correction of C/N for the insertion of a preamplifier is simple, although the equations and graphs seem to be complicated. The preamplifier reduces the C/N of the cable system because of its own noise contribution. This shows up as a reduced C/N at the preamplifier output. And here's the catch: The amount of C/N reduction requires that you know the corrected carrier-to-noise ratio at the preamplifier output. So all the corrections discussed so far must be applied to the spectrum analyzer's reading before compensating for the preamplifier C/N error.

TIP

A preamplifier is an active device that adds its own noise. It makes C/N look worse than it is to the spectrum analyzer.

Correcting for Preamplifier Noise Is Just a Graph Look-Up

The good news is that the process of correcting for preamplifier noise is an easy procedure. Keep in mind that the C/N of the system is always better than the measured C/N. That is, the analyzer measures the worst-case C/N. If this value is sufficiently better than the performance required for compliance, then no further corrections need be applied. However, with most system testing, accurate data is maintained in order to see trends in overall performance so that preventative maintenance can be done before the system goes out of specification.

Preamplifier correction requires these inputs:

- ✦ C/N at the preamplifier output corrected for noise power bandwidth, 4 MHz bandwidth, resolution bandwidth, noise-near-noise, and log detection.
- ✦ An accurate measurement of the preamplifier's gain at the channel frequency to be tested.
- ✦ The carrier level at the preamplifier's output.

All these are readily derived from the testing done so far. It is good practice to confirm the manufacturer's preamplifier gain, especially if the preamplifier's specified gain tolerance is more than ±1 dB.24

The correction to C/N is simply a matter of looking up the value from a graph. Figures 67 and 68 represent different ranges. Here is an example of determining preamplifier C/N correction with the graphs.

Example 33. Correcting C/N for a Preamplifier

A 7 dB noise figure preamplifier with a gain of 21 dB is used to boost system noise high enough for the spectrum analyzer to see system noise 10 dB higher than the spectrum analyzer noise. Measured C/N is 71.2 dB with the carrier at +22.6 dBmV, measured at the preamplifier output. The corrected C/N is 71.2 − 21.25 − 2.5 dB = 47.45 dB. What is the system C/N corrected for the preamplifier?

The horizontal axis in Figure 67 is a combination of carrier level, preamplifier gain and the corrected C/N. This is +22.6 dBmV − 21 dB − 47.45 dB = −45.85 dB. Find

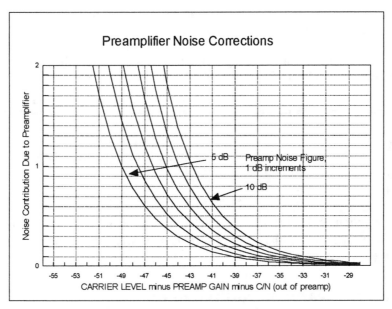

Figure 67. Corrections to C/N when adding a preamplifier.

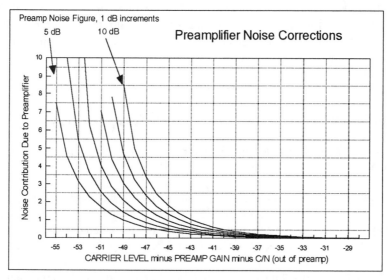

Figure 68. Detail Corrections to C/N when adding a preamplifier.

−46.3 dB along the horizontal axis and read up to the 7 dB preamplifier noise figure line. The C/N correction of 1.1 dB is read from the vertical axis.

The system carrier-to-noise is always better than at the output of the preamplifier, so the system C/N is 47.9 dB + 1.1 dB = 49 dB.

TIP	**The preamplifier is just a simple look-up value. The correction improves the measured C/N.**

The noise correction could have been calculated from the equation given by an exact equation, but the accuracy of the graph is sufficient, as you will see in the discussion of accuracy at the end of this chapter.

System Noise-Measurement Procedure

This section takes you step-by-step through two carrier-to-noise measurements. The first is a measurement of a channel whose modulation can be removed, and complies with the FCC regulations. The second is a noninterfering test, which makes the measurement on the upper channel sideband while the modulation is on. Although the second procedure does not conform strictly to compliance regulations, it is practical, speedy, and a good indication of system noise performance.

Example 34. Measure the carrier-to-noise to FCC compliance rules.

Measure the C/N of a channel for which the modulation can be removed during the measurement.

With the carrier modulation off, tune to a frequency between the visual carrier and the aural carrier, being careful to avoid areas where coherent signals are present. Set the analyzer to a resolution bandwidth of 30 kHz and the video bandwidth to 100 Hz. Disconnect the input to the spectrum analyzer. The noise drop, N_{DROP}, is zero.

Add a preamplifier with a noise figure of 9 dB and a gain of 21 dB to the input of the spectrum analyzer. Test for overload by using the same settings as for carrier level,

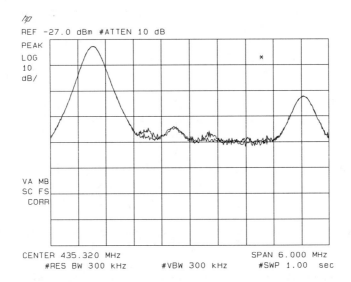

Figure 69. Overload test for an unmodulated carrier.

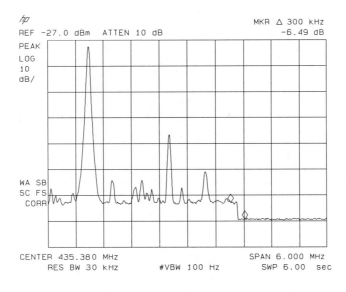

Figure 70. Noise drop with the disconnect test.

except monitor the signal peak with marker or two traces as in Figure 69. Use one trace to store a single one-second sweep at one attenuation setting, 10 dB in this case. Then change traces and repeat with an attenuator setting of 20 dB. The figure shows no change in signal amplitude, and therefore no overload. Return the analyzer attenuator to 10 dB.

Repeat the disconnect test by sweeping a single time over the span, and disconnecting the input to the preamplifier. The resulting noise drop can be measured with markers as shown in Figure 70. This results of this disconnect test shows a N_{DROP} of 6.5 dB. Note that neither noise signal is within the lowest graticule of the display. This lowest division is often uncalibrated, as it is in this specific analyzer. If the noise ventures into this area, simply decrease the reference level until the noise is within the calibrated display range and repeat the disconnect test.

Look up the noise-near-noise correction, C_{NNN}, from Figure 62. For this example $C_{NNN} = 1.0$ dB.

With the modulation on, tune to the carrier and set the analyzer resolution and video bandwidths to 300 kHz. Bring the carrier to the reference level, and use trace maximum or slower sweep times to ensure the peak is at the top. Record the carrier level as C_{LEVEL} in dBmV. In this measurement $C_{LEVEL} = +23.5$ dBmV. Remember that C_{LEVEL} is a measure after the preamplifier, and is not the true carrier level. Since both the carrier and noise are subject to the same amplifier gain, the C/N, which is a ratio, is accurate. Just don't confuse this carrier level with the actual system level.

Once again, remove the modulation and measure the noise level within the channel boundary, that is, anywhere from the carrier to 4 MHz above the carrier. Set the detector to the sample mode, the resolution bandwidth to 30 kHz, and video bandwidth to 100 Hz. Take a sweep and then use the marker to measure a spot free of coherent disturbances, as Figure 71 demonstrates. The uncorrected noise level, N_{RAW}, is −43.75 dBmV in a 30 kHz bandwidth.

Calculate the corrected noise level, $N_{RAW} + 2.5 + 21.25 - 0.51 - C_{NNN}$. where 2.5 dB is the correction for log detection, 21.25 dB is the correction from 30 kHz to 4 MHz resolution bandwidths, and 0.51 dB is the correction for the noise power bandwidth. $N_{CORR} = -43.75 + 2.5 + 21.25 - 0.51 - 1.0. = -21.51$ dBmV/ 4 MHz.

The C/N is the difference between the carrier level and the corrected noise level, C/N $= C_{LEVEL} - N_{CORR} = +23.5$ dBmV − (−21.51 dBmV/4 MHz) = 45.01 dB.

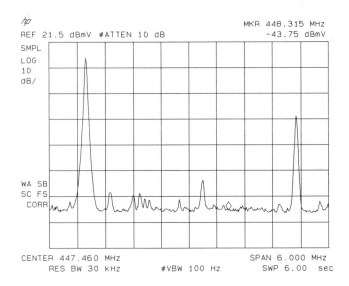

Figure 71. Measuring the noise level within the FCC boundary.

Keep in mind the logic of the signs of each of the correction factors. The log detection factor always increases the noise level because the analyzer's log and detection squash the display of noise, showing it lower than actual. The correction for bandwidths is in the same direction, that is, adding to the noise level, because the level is supposed to be for a much wider bandwidth than the analyzer is capable of using. And wider bandwidth means the correction adds to the noise level. The correction for noise power is lower because Gaussian filters have a wider equivalent bandwidth than the 3 dB filter bandwidth; thus the analyzer "thinks" it is measuring in 30 kHz but is actually measuring with a filter that is slightly wider, and noise power must be decreased to compensate.

Channels or pilots that are not processed through modulators, that is, they have had no video signal conversion and/or amplification, are not good C/N test signals. Their noise represents the system noise, but may not represent channel noise. C/N must be measured on a carrier whose modulation has been turned off at its origin, but the carrier level must be taken from either the same carrier modulated or from an adjacent active channel with the assumption that the system flatness varies <1 dB between the active carrier and the noise measurement frequency.

One more correction needs to be made. A preamplifier was added to boost system noise up to where the spectrum analyzer could measure it. The preamplifier noise reduces the C/N seen by the analyzer. Continue with the above example to correct for the preamplifier. Assume that the preamplifier has a gain of 21 dB and a noise figure of 9 dB.

Correction for the preamplifier requires the C_{LEVEL}, the corrected C/N, and the gain and noise figure of the preamplifier. Using Figure 67, find the horizontal point as C_{LEVEL} – gain – C/N = +23.5 dBmV – 21 dB – 45.01 dB = –42. 51 dB. From this value on the horizontal axis of Figure 68, look up to the 9 dB noise figure line to find the correction of 0.7 dB.

The system carrier-to-noise is larger than at the preamplifier output. C/N = 45.01 + 0.7 = 45.71 dB.

A quick look at Figures 67 and 68 tells you that for high carrier levels in systems with moderate-to-low C/N, the preamplifier does not add significant error. If the carrier levels are low, and/or C/N is high, the preamplifier has significant role to play in the accuracy of the measurement.

Quick C/N Measurements

Quick field measurements are essential for spot checking and troubleshooting. It is not convenient or possible to turn off the modulation to a channel during this testing, so another technique has been widely accepted practice: measuring C/N 2 MHz below the channel carrier. The modulation interferes with the method, undoubtedly, but the results can be assumed the best of the worst case; that is, if the C/N is at an acceptable level, the system is probably performing better. The next example demonstrates how to get a quick C/N measurement when channel modulation is on.

TIP

You can make a quick estimate of C/N by measuring the noise power near a modulated channel.

Example 35. Measure C/N with modulation on.

Measure the C/N of a modulated channel using the setup in Example 34.

The modulation below the channel carrier frequency is reduced by vestigial filtering. If there is no lower adjacent channel, such as the case for channels 2, 4, and 7 in a standard tuning configuration, then a spot 2 MHz below the carrier can be used.

Tune to one of these channels, placing the carrier in the center of the display with the span set to 6 MHz and the resolution bandwidth at 300 kHz. Set the carrier peak at the top graticule and read the carrier level as shown in Figure 72. The carrier is +36.0 dBmV.

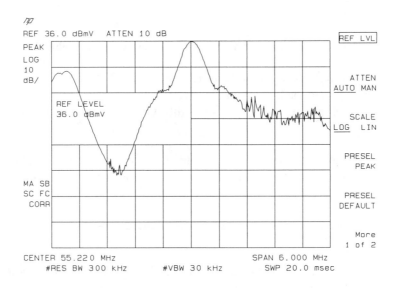

Figure 72. Carrier measurement.

Without changing the reference level, change resolution bandwidth to 30 kHz and the video bandwidth to 100 Hz. Place a marker about 2 MHz below the carrier, being careful to avoid any composite distortions signals, the small CW spikes, as shown in Figure 73. Place a marker on the noise level where it is flat. If the marker is in the lowest division, decrease the reference level to bring the noise into the second division from the bottom. Set the detector to sample mode.

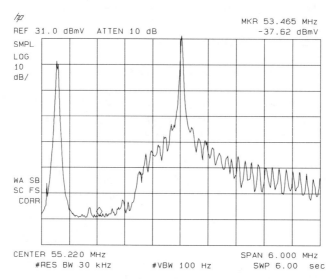

Figure 73. Noise measurement on the lower vestigial sideband.

Read the uncorrected noise from the marker, N_{RAW}, as −37.62 dBmV/30 kHz. Apply the corrections as in the last example to give N_{CORR} = −37.62 + 2.5 + 21.25 − 0.51 − 1.0. = −15.38 dBmV/ 4 MHz. C/N = +36.0 dBmV − (−15.38 dBmV/4 MHz) = 51.38 dB.

Since modulation from the visual carrier and the lower adjacent service pilot adds to the noise reading, the C/N is a worst-case value, but still much better than the requirement.

If the C/N measured with this technique is close to the regulation values, additional testing using more formal procedures is required. However, as long as the C/N levels are sufficiently above the regulation values, you can even measure directly between adjacent carriers as shown in Figure 74, where the dip in noise between the visual carrier and adjacent audio carrier provides a signal-free noise floor.

TIP   **Measuring C/N with modulation on is a worst-case value. If this value meets FCC regulations, you need measure no further. If it is close, do a more formal measurement.**

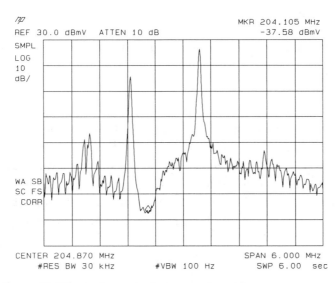

Figure 74. Measuring noise between channels.

The same preamplifier corrections would apply to this measurement as it did in the compliance example before.

Accuracy of C/N Measurements

Uncertainties are not to be confused with error corrections. Uncertainties limit the measurement accuracy. Corrections are known values applied for a correct answer. An example of a correction is the conversion from 30 kHz to 4 MHz bandwidths when evaluating noise-power density. This is a definite mathematical relationship, which is as accurate as the calculator making the computation. Another correction example is the noise-power bandwidth correction. It too is a definite mathematical relationship, but the data fed into the equation depends upon how carefully the integration of the power under the analyzer's resolution bandwidth filter was collected.

TIP

Remember that corrections are not uncertainties; they just correct for a known error.

Uncertainties offer the most trouble with C/N. Corrections are fairly well behaved provided you have accurate and up-to-date calibration data on your analyzer, and its manufacturer has provided accurate correction factors, such as log detection correction.

Amplitude Uncertainties

The uncertainties for amplitudes of carrier and noise are the same as introduced in Chapter 4. The carrier-to-noise measurements, although treated like two absolute measurements, have the uncertainties of relative measurements. The uncertainty when comparing the carrier level, set to the reference level is only the display fidelity, a value less than ±0.75 dB. If, after the carrier level was measured, the reference level is changed to bring the noise out of the last division, another ±0.3 dB must be added to account for new reference level accuracy. See the example spectrum analyzer specifications in Appendix C.

Amplitude Uncertainties That Can Be Ignored

Flatness does not contribute to amplitude uncertainty for C/N because the measurement is made at essentially one frequency. Calibrator amplitude accuracy and the addition of a preamplifier are also excluded because they are absolute amplitude uncertainties.

But don't change the input attenuator during measurement. The input attenuator, being a wide band RF device, has ±0.5 dB switching uncertainty. Fortunately, normal measurements of C/N do not require changing the attenuator between the carrier and noise measurements.

TIP

Don't change the analyzer's input attenuator between measuring the noise and the carrier. Its switching uncertainty can be high.

Errors Not Subject to Uncertainty

Errors are not uncertainties, but they are often computed corrections based on measured data, and, therefore, could be subject to uncertainty. Because some error corrections are provided by the manufacturer of the test equipment, they are accepted as exact. Log detection and noise-near-noise corrections are such corrections in this category. With noise-near-noise, follow the procedures in this book to ensure the most accurate disconnect noise difference value. And don't forget to terminate the analyzer or preamplifier input with 75 Ω when measuring its noise. Certainly the derivation of the noise-near-noise graph and the log detection correction is beyond scope of this book. Rely on the spectrum analyzer manufacturer and assume the corrections are free from uncertainty.

Another set of errors not subject to uncertainty on face value are those that correct for measurement bandwidths of the analyzer, such as from resolution bandwidth to 4 MHz, or resolution bandwidth to noise-power bandwidth. These are simply mathematical expressions, dependent upon how many decimal places your calculator can yield.

NEPBW Includes Resolution Bandwidth Uncertainties

The spectrum analyzer resolution of 3 or 6 dB bandwidths is generally only accurate to ±15%. Fortunately the computation of NEPBW with the techniques taught in this chapter eliminate the uncertainty due to sloppy resolution bandwidth shape because of the physical integration of the power under the bandwidth.

Preamplifier Gain and Noise Figure

From the figures and examples in the preamplifier correction section, you see that the preamplifier always makes C/N look worse. If the preamplifier correction is low, <1 dB for example, for your usual measurements, it can be included in your measurement guardband or ignored and used as an unseen safety margin. The uncertainty of the lower correction values is also minimal. However, above 1 dB the preamplifier correction is subject to much higher uncertainties. These can be seen in Figures 67 and 68 where the curves are steep. A ±1 dB preamplifier gain misreading can cause as much as ±3 dB correction uncertainty. This uncertainty affects C/N accuracy dB for dB.

Other Uncertainties

The uncertainties and corrections caused by the system noise being other than random, as assumed for this work, are not covered here. But as noted in the last examples, you can improve accuracy by avoiding non-noise signals, such as coherent disturbances and modulation sidebands, when setting the marker on the noise

Computation of C/N Accuracy

The following example illustrates C/N accuracy.

Example 36. Calculate C/N accuracy.

How accurate is the C/N of 45.01 dB in Example 34?

As in other accuracy considerations, the test procedure is the key. The carrier was measured by bringing the carrier to the reference level, but then the reference level was changed, so both display scale fidelity and reference level accuracy apply. These are ±0.75 and ±0.3 dB, respectively. They add directly for a total uncertainty of ±1.05 dB. The C/N is 45.01 ± 1.05 dB, or a value between 46.09 and 43.93 dB.

If the system specification for C/N is 43 dB, a guardband guarantees that the system C/N is always better, even with an uncertainty of ±1.05 dB. The guardband is 43 + 1.05 dB = 44.05 dB, or rounding to 44.1 dB. Thus, any C/N better than 44.1 dB, is insurance against actual C/N below 43 dB.

TIP	**Typical C/N uncertainty is about ±1.0 dB if you follow the guidelines in this chapter.**

Summary of C/N Measurement Procedures

Here are the major lessons to be learned in this chapter:

- Measurement of the visual carrier with 300 kHz resolution bandwidth.
- Use of the disconnect test to determine whether a preamplifier is required.
- Use of a preamplifier with <10 dB noise figure and between 20 and 30 dB gain.
- Tests for overload, adding a bandpass filter if necessary.
- Repetition of the disconnect test after adding the preamplifier and/or filter.
- Use of the sample detection mode, 30 kHz resolution bandwidth, and 100 Hz video bandwidth for measuring noise.
- More accurate measurements if the reference level is not changed between carrier and noise measurements.
- Avoidance of changing the attenuator between carrier and noise measurements.

- Corrections to the noise level for 4 MHz bandwidth, resolution bandwidth, IF noise equivalent bandwidth, log detector, and noise-near-noise
- Use of a preamplifier, if disconnect test shows <3 dB change. It degrades the measured C/N by an amount dependent upon the preamplifier gain, noise figure, carrier level, and corrected C/N. The error can be high when measuring low carrier levels in systems where the C/N is high.
- Estimation of the C/N accuracy to provide a guardband for assuring specified system performance.

Selected Bibliography

- Bullinger, Rex. "How to Measure Carrier-to-Noise," *Communication Engineering and Design*, November, 1994.
- Engelson, Morris. *Modern Spectrum Analyzer Measurements*. Portland: published by JMS, 1991.
- *Fundamentals of RF and Microwave Noise Figure Measurements*. Hewlett-Packard Company, Application Note AN 57-1, Literature No. 5952-8255, Palo Alto, CA, July 1983.
- *NCTA Recommended Practices For Measurements on Cable Television Systems*. 2nd ed., October, 1993.
- Peterson, Blake. *Spectrum Analysis Basics*. Hewlett-Packard Company, Application Note AN 150, Literature No. 5952-0292, Santa Rosa, CA, 1989.

8

Coherent Disturbances - Beating CSO and CTB

Overview

Interference at the customer receiver can be caused by signals generated within the cable system, by signals that are received at the head end, or by signals that intrude into the system along its distribution path. The most troublesome distortion generated by systems carrying over 24 channels is that created by visual carriers beating against one another to form unwanted interference signals called composite triple beat, CTB, and composite second order, CSO. These signals are most disturbing because they lay within the transmitted channel bandwidths, impossible to filter or correct farther down the cable system. The distortion produced is a chief cause of subscriber complaints because of its visually annoying nature. This chapter provides an understanding of how CSO/CTB produces irate calls from your subscribers, and how you can measure these beats with a spectrum analyzer long before they become troublesome.

Coherent Disturbances

The measurements for CSO/CTB are assumed to be at the subscriber tap, not at the output of set-top converter.

Why the Subscriber Complains about CSO/CTB Beats

CSO/CTBs are interference signals formed from the mixing of carriers. When these interference signals fall into the video range of a channel, they are indistinguishable from the channel modulation, and appear as a beat pattern or as lines on the picture of the television receiver. The following figures simulate the effects of CSO/CTB on the television picture.

Figure 75. An undistorted, simulated television picture.

Compare Figure 75 to Figure 77. CSO, as simulated in the second figure, is particularly annoying because it looks like the picture needs fine tuning. But even so-called "fine-tuning" on older TV receivers would not eliminate CSO; it is an integral part of the incoming signal, not the result of adjacent channel interference.

The Measurement in Brief

Before launching into the details, here is how to measure CSO/CTB when control over the channel under test modulation and carrier can be turned off.

1. Measure the channel carrier by following the guidelines in Chapter 4. Set the carrier level to the analyzer's reference level. Record the level in dBmV.
2. Set the spectrum analyzer span to 6 MHz, resolution bandwidth to 30 kHz, video bandwidth to 1 kHz, and use a digital video averaging with a samples set to 5. If your analyzer does not have digital video averaging, set the video bandwidth to 30 Hz. Set the analyzer's detector to the sample mode.
3. Have the channel modulation turned off and locate the CSO beats 0.75 and 1.25 MHz above the carrier.
4. Measure the highest CSO level using a marker, and test for overload by inserting a 2 or 3 dB attenuator at the analyzer (or preamplifier) input. The CSO value should change only by the amount of the attenuation. If distortion is present, add a bandpass filter and retest.
5. Record the CSO level.
6. Perform the disconnect test as in Chapter 7. If the noise changes less than 10 dB, apply a correction, in dB, to the CSO level. Remember that CSO appears on the analyzer higher than it really is, so the corrected CSO is smaller.
7. Calculate the carrier to CSO ratio, or C/CSO, as the difference between the carrier and CSO levels.
8. Have the carrier turned off and measure CTB at the carrier level. CTB is measured the same was CSO is measured, corection and all.

The horizontal streaks of noiselike interference that are common symptoms of CSO, simulated in Figure 77, are more like the interference from automobile ignitions when the television receiver is receiving off-air from an antenna.

Figure 76. Composite triple beat that often appears as horizontal streaks covering one or more lines of video.

Figure 77. Composite second-order distortion that usually appears as swimming diagonal stripes in the TV picture.

FCC Regulations for CSO/CTB

Here are the FCC regulations governing the specifications for and testing of CSO/CTB.

Regulation: FCC 76.605 (a) (8) and 76.609 (f)

Regulation Text For Technical Standards: "The ratio of visual signal level to rms amplitude of any coherent disturbances such as intermodulation products, second- and third-order distortions or discrete-frequency interfering signals not operating on proper offset assignments shall be: (i) the ratio of visual signal level to coherent disturbances shall not be less than 51 dB for noncoherent channel systems, when measured with modulated carriers and time averaged; and (ii) the ratio of visual signal level to coherent disturbances which are frequency-coincident with the visual carrier shall not be less than 47 dB for coherent systems when measured with modulated carriers and time averaged."

Regulation Text For Measurements: "The amplitude of discrete frequency interfering signals within a cable television channel may be determined with either a spectrum analyzer or with a frequency-selective voltmeter (field strength meter), which instruments have been calibrated for adequate accuracy. If calibration accuracy is in doubt, measurements may be referenced to a calibrated signal generator, or a calibrated variable attenuator, substituted at the point of measurement. If an amplifier is used between the subscriber terminal and the measuring instrument, appropriate corrections must be made to account for its gain."

What This Means, Practically Speaking

Here are the regulations in abbreviated terms.

- CSO/CTB distortion products must be more than 51 dB below the visual carrier level in systems where the carrier frequencies are not tied to a single synchronizing generator.
- CSO/CTB distortion products for IRC systems need be 47 dB down.
- The distortion signal level must be averaged over time.
- A spectrum analyzer can be used to make the measurement.
- Measurements may require removing modulation of the carrier from the channel under test.

It has been shown that CTB can be perceived with as little distortion as 57 dB under ideal program and receiver conditions. The spectrum analyzer, with its every wide amplitude measurement range, is ideal for seeking out and measuring these beats.

Why So Many Beats?

As mentioned, CSO/CTBs are the result of the distortion of the signals being transmitted through the cable system. Distortion is defined as any undesired change in the waveform of a signal in the course of its passage through the cable system. Distortion usually manifests itself as signals, called beats, added to the cable transmission. When these signals fall within a channel bandwidth, they cause interference with the quality of channel reception. When the distortion signals fall outside the program material bandwidth they cause no distortion, and, in fact, are useful in determining the general beat level present in the system.

Composite Distortion

Cable system CSO/CTB is not one single distortion signal at one exact CW frequency. It is a collection of many distortion signals falling on top or near enough to one another to be indistinguishable by the spectrum analyzer or signal level meter. The effect of having several signals measured with a bandwidth wide enough to see their collective amplitude makes the signal a composite. Composite means the total signal amplitude due to a number of distortion signals packed into a specified frequency span.

Beats by the Numbers

As cable systems add additional channels, distortion from cascaded amplifiers increases. Systems with 13 to 20 channels have little CSO or CTB distortion. In these systems, the most troublesome interference is cross modulation, the subject of the following chapter. Cross modulation does not increase as fast as CSO/CTB with an increase of the number of channels. With CSO/CTB the number of interference signals increases as the square of the number of system channels. For example, in the center of a system containing 20 channels, the number of CTB distortion signals generated at that channel's frequency is about 150. The CTB number of signals at the center frequency of a 60-channel system is greater than 1,350! This is illustrated in Figure 78. The number of CTB signals at channels at the edges of the system are about two thirds the number than those channels at the center frequencies. This is an important characteristic of CTB to remember when it comes time to set up a test procedure.

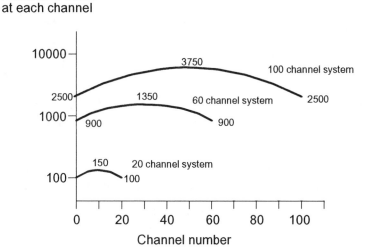

Figure 78. Quantity of CTB signals generated in systems with different numbers of channels.

Discrete Distortion Defined

Before tackling the composite nature of the CSO/CTB signals, let's understand how distortion products are generated. A discrete distortion product or beat is defined as one whose source can be traced to a single set of carriers. These discrete distortion products are very small beats in cable systems because system amplifiers have very low distortion levels. Therefore, discrete distortion products are very low. But it is a healthy exercise to study how discrete harmonic and third order distortion is generated as a background to understanding composite beats and the way a spectrum analyzer can, and sometimes, cannot make the measurement.

Harmonic Distortion

Any time signals pass through a nonlinear device, whether the device amplifies, attenuates, or changes the frequency of the input signal, the input signal energy is disbursed to signals other than the exact desired output. For example, in an amplifier the distortion may show up as harmonic signals, so-called because their frequency values are related by an integer such as 2, 3, or 4 times the input signal frequency. Figure 79 shows this relationship between the input and output frequencies of an amplifier.

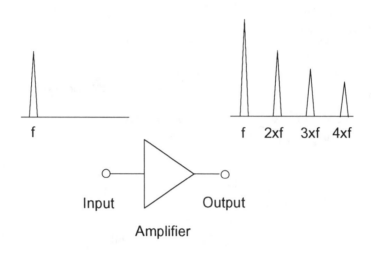

Figure 79. Harmonic distortion created by an amplifier.

The spectrum analyzer views discrete harmonic distortion just as it looks at other types of signals. Here is an example:

Example 37. Measure discrete harmonics with the spectrum analyzer.

Measure the relative strength of the analyzer's calibration signal's second harmonic.

Connect the calibration signal to the analyzer's input and set the span and reference level to see the first several harmonics along with the signal. Figure 80 shows this. Simple harmonic distortion is usually measured by comparing the amplitude of the carrier, or fundamental, to the amplitude of the strongest harmonic or harmonics. In this example, the fundamental at 300 MHz is compared to its second harmonic at 600 MHz. Place the marker at the peak of the fundamental, and the second marker at the next signal to the right, the second harmonic. This harmonic is 23.64 dB down from the fundamental.

These harmonic distortion signals are called discrete because they are the direct result of a single signal's distortion. CSO is the result of harmonic distortion built up as signals are

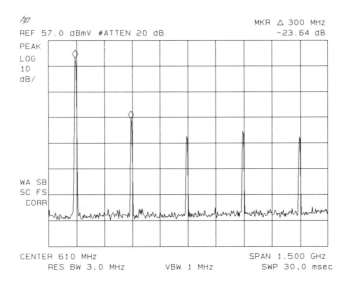

Figure 80. Spectrum analyzer's calibration signal and its harmonics.

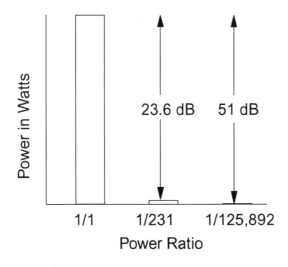

Figure 81. Comparison of the power in beats.

amplified through the cascaded trunk and distribution amplifiers, with literally hundreds of harmonic signals mixing with one another to produce beats at very low power levels.

Another typical harmonic distortion characteristic can be seen in Figure 80. The higher the harmonic distortion frequency, the lower the distortion usually is. There are exceptions, of course, but the tendency is for less harmonic power at the higher harmonic numbers. This is why harmonics above the second or third levels are not considered in the distortion computations. The higher levels are usually so much lower in power than the fundamental, second, and third that their effects can be ignored. As an example, look at the bar graph in Figure 81. The bar heights are proportional to the power level in watts. The carrier of Example 37 is compared to the power of the harmonic, which is 23.6 dB lower. The power difference is $\log^{-1}(23.6/10)$, or 231 times smaller than the fundamental. The third bar represents the level of a CSO beat at the FCC-specified 51 dB below the carrier. This beat is $\log^{-1}(51/10)$, or 125,892 times smaller than the fundamental!

Discrete Third-Order Distortion

The source of CTB, or composite triple beat, is third-order distortion. Third-order means that three signals have mixed in a nonlinear device to produce a distortion signal. If the three signals' frequencies are represented by A, B, and C, then the frequencies of the third-order distortions are given by the addition and subtraction of any combination of A, B, and C, including the second harmonics of each carrier. The second harmonics are noted as 2A, 2B, or 2C, another way of saying +A+A, +B+B, or +C+C. The ± means + or −, so each term will have a number of values, depending upon how many combinations of ± terms there are.

- ±A ±B ±C
- ±2A ± B
- ±2A ± C
- ±2B ± A
- ±2B ± C
- ±2C ± A
- ±2C ± B

As you can see, there are practically an unlimited number of beats possible. But the signals that are a problem for the cable system are only those that lie within the cable system frequency range. An example illustrates which of these many distortion products pose a problem.

Example 38. Discrete third-order distortion.

Calculate all the third-order distortion products from carriers at 50, 200, and 325 MHz. Assuming that the cable system operates from 50 to 450 MHz, which of the products interfere with the other system channels?

Set A = 50 MHz, B = 200 MHz, and C = 325 MHz. The first observation is that all the signals' third-order beats with + terms will lie above the highest signal, and therefore need not be calculated. Similarly, all the second harmonic products with the third signal added are outside the cable frequency range. The terms with two times the highest value, C, can also be thrown out because the frequencies lie above C. The remaining terms need to be calculated to see if they are in the system range. These are listed.

Term	Frequency (MHz)
C–A+B	475
C+A–B	175
B–2A	100
C–2A	225
2B–A	350
2B–C	75
2C–A	600
2C–B	450

Figure 82 shows the distortion products plotted in relation to the three carriers. Five of the products lie within the cable transmission range.

When another carrier is introduced to the simple signal set of Figure 82, the number of third-order distortion products increases by more than ten times. But more important, some of the new distortion products will fall on top of the distortion already created by the first three signals. For example, if the new carrier is at 425 MHz, the third-order distortion produced by +2(325 MHz) – 425 MHz, would fall at 225 MHz, right on top of the +C–2A product in the above example. In a system with just 12 carriers, a computer or a pot of coffee and sharp pencil would be required to calculate all the possible combinations of CTB that can fall within the cable frequency band.

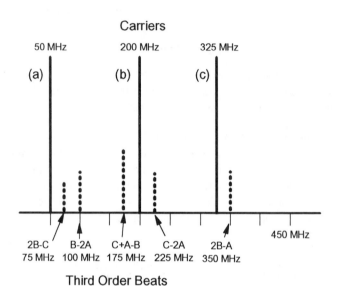

Figure 82. Third-order distortion products from three carriers.

Composite CSO/CTB – the Fuzzy Signal

In a standard frequency allocation cable system the carriers are not locked to one oscillator and therefore are free to roam slightly within a few hundred hertz depending upon time, temperature, and equipment conditions. These slight variations in each signal frequency mean that distortion products will not lie directly upon one another. In addition, the horizontal sync pulses, whose amplitude determine the visual carrier's level, are not in phase. So even when distortion products are at exactly the same frequency, it is only occasionally that the sync pulses occur together. Recall, however, that the number of CTB distortion products for a 60-channel system can be as high as 1,350. So, what the distortion beats lack in power, they more than make up for in quantity.

Composite Beat Is Averaged Like Noise

The result of all the beats falling within a small frequency span and bringing along their own slight variations in frequency and sync phase, is a composite signal that behaves like avery noisy CW signal. The signal amplitude is always changing. So, how is it measured? In the spectrum analyzer, the variations in frequency are tackled by setting the resolution bandwidth wider than the expected frequency variation. The amplitude variations caused by phasing in and out of all the composite contributors are averaged by the analyzer's video bandwidth. To see how this works see Figure 83. This beat resides above all the channels in a 52-channel system, and has been exaggerated by using two stages of amplification in front of the spectrum analyzer.

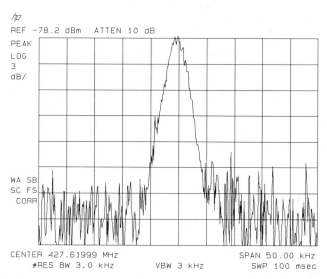

Figure 83. A single distortion product outside the regular channel span.

Although it appears as any CW signal, the amplitude variation riding on the signal's maximum response appears random, that is, not the result of any recognized television modulation format, such as periodic sync pulses. The random noise is not coming from the spectrum analyzer; the signal is >10 dB higher than the noise floor. In fact this random modulation is inspected closer using the spectrum analyzer as a fixed tuned receiver, and looking at the demodulated video of the signal, which is similar to the view a waveform analyzer might present. Figure 84 shows the time domain demodulated waveform in 600 μsec full span. There is no sign of horizontal sync pulses, demonstrating the random, or composite, nature of this beat.

| TIP | If the signal you are inspecting does not seem to change amplitude over time in a random way, it may not be a composite signal. It could be ingress. |

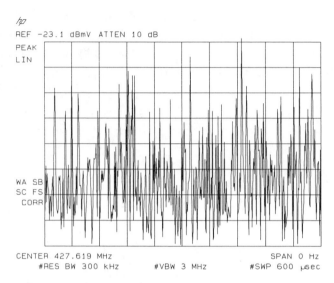

REF -23.1 dBmV ATTEN 10 dB

PEAK
LIN

WA SB
SC FS
CORR

CENTER 427.619 MHz   SPAN 0 Hz
#RES BW 300 kHz   #VBW 3 MHz   #SWP 600 μsec

Figure 84. Distortion beat of Figure 83 demodulated as a waveform.

Taming the Beats

So, if the beat measurement from CSO or CTB is subject to such amplitude variations, how is a consistent measurement made? Simple. Widen the resolution bandwidth to 30 kHz and average the amplitude variations by narrowing the analyzer's video filter. This effect is seen in Figure 85. As with any noiselike signal, the analyzer's detector should be set to the random, or sample, mode so that the signal's random amplitude variations are not skewed toward the higher peaks. Figure 86 shows the same span with the video bandwidth set to 30 Hz and the detector set to sample mode.

The response is smoothed for a repeatable amplitude measurement of the changes that occur over short periods of time–milliseconds to seconds. Beats may have longer periodic changes. These will be resolved in the measurement section later in this chapter.

TIP

Theoretically, the highest concentration of CSO is at channels 4 and 5. For CTB, the largest grouping is just above the middle of the frequency span.

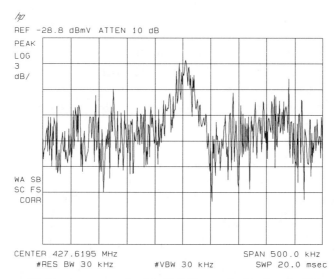

Figure 85. Beat measured with 30 kHz resolution bandwidth and wide video bandwidth.

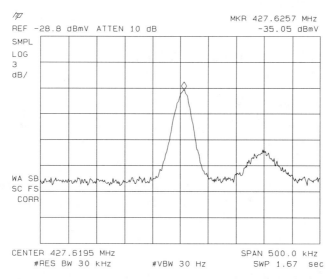

Figure 86. Beat measured with 30 kHz resolution bandwidth, 30 Hz video bandwidth, and sample detection.

Where Are the Beats Hiding?

In the standard frequency allocation cable systems, the worst case, that is, the highest CSO/CTB distortions, are located in well-known places on and near the visual carriers. For CTB, the worst distortion is exactly at the carrier frequency. For CSO, the worst distortion products are found ±0.75 MHz and ±1.25 MHz from the carrier frequencies. Figure 87 shows this graphically.

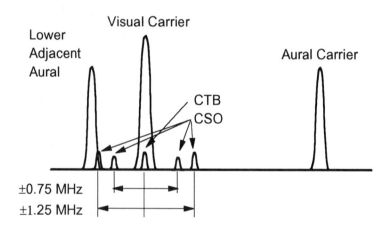

Figure 87. CSO/CTB distortion frequencies.

The CSO distortion products that fall on the lower side of the carrier do not cause interference because they are outside the channel bandwidth, but provide a simple way to measure CSO when carrier modulation is on.

TIP **Out-of-channel CSO and CTB measurement is a quick way to determine your system's performance without having to interrupt programming.**

Calculating CTB Frequencies

You may ask yourself, why does CTB occur where it does? Let's look at the dominant CTB products first. The fundamental carriers are the major source of energy for distortion in third-order intermodulation. And most of these products are generated in the form of A + B − C, where the letters represent channel carrier frequencies, and where A + B > C. This distortion frequency will always land on a channel carrier, whether or not it is occupied. As an example, take the frequencies of channel 21 for A, channel 8 for B, and channel 11 for C. See Figure 88(a). The distortion will appear at 163.25 MHz + 181.25 MHz − 199.25 MHz = 145.25 MHz, which is the carrier frequency for channel 18. Try this on your own with practically any three channels, it always seems to work.

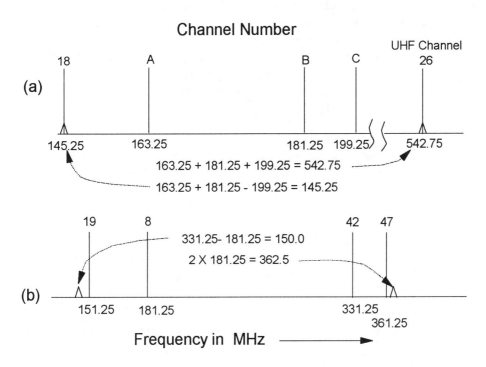

Figure 88. Beats resulting from (a) composite triple beat and (b) composite second-order mixing of channel carriers.

The fundamental carrier frequencies can also add as +A + B + C. The distortion will land in the middle of a fourth channel's video bandwidth. Channels 2, 3, and 4 carrier frequencies add to a distortion frequency of 183.75 MHz. This signal is at 2.5 MHz above the channel carrier of channel 8, whose carrier is at 181.25 MHz. The +A + B + C values mostly lie above 500 MHz so they comprise a small portion of the total interfering CTB signals produced in standard frequency systems.

CTBs fall right on the visual carrier frequencies. They also can be found on 6 MHz center multiples outside the channel boundaries, such as between channels 4 and 5.

CSO Frequencies

CSO in standard frequency allocation systems are found at ±0.75 and ±1.25 MHz from visual carrier. To see why these beats appear where they do, take the frequencies of any two channels from the middle and upper channel bands and combine them to produce a second-order distortion beat frequency. For example, subtract channel 8 at 181.25 MHz from channel 64 at 331.25 MHz. The result is 150.0 MHz. This beat is 1.25 MHz below the channel 19 carrier at 151.25 MHz. See Figure 88(b). The channel 8 second harmonic, at 2×181.25 MHz, is at +1.25 MHz above the channel 47 carrier.

CSO beats are above and below the visual carrier 0.75 and 1.25 MHz. The lower 1.25 MHz beat can often be measured between modulated channels.

The CSO beats at ±0.75 MHz are the result of standard allocation channels 5 and 6, which are offset 2 MHz from 6 MHz multiples, mixing with the other channels. It is not obvious why this shift causes beats at ±0.75 MHz, so here is an example. Add channel 6 to channel 8 to get 77.25 MHz + 181.25 MHz = 258.0 MHz. This is 0.75 MHz below the channel 30 carrier at 259.25 MHz. Because only two of the channels are offset in the standard frequency system, the ±0.75 MHz CSO is generally lower in amplitude than the ±1.25 MHz beats.

In IRC allocation systems the CSO will fall exactly ±1.25 MHz around the carrier, with no ±0.75 MHz beats. In HRC systems, both CSO and CTB fall exactly on the carrier frequency.

Where They Fall

There may be areas where CTB or CSO are worse than others, but because of the complexity of cable systems today, there are no guidelines that would stand up to close scrutiny. The distribution of distortion beats over the system frequency range may help you decide where to monitor to detect performance changes before they get bad enough to draw customer complaints.

The center channels are burdened with about 30% more CTB distortion signals than are the low and high channels, as shown in Figure 78. For CSO the burden shifts from the center of the system to the lowest and highest channels, as illustrated in Figure 89. But given the composite nature of the distortion signals and the variety of cable plant designs, the measured levels may or may not follow these beat population curves.

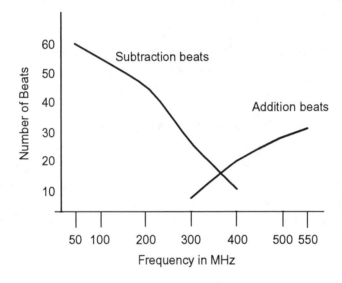

Figure 89. Mathematically derived population of CSO beats in a 77-channel system.

Increases in CSO levels at the destination hubs of fiber and AML paths have been observed. The reason may be the nonsymetrical distortion characteristics of amplification associated with the modulation and demodulation of the channel groups.

Practical Measurement Procedures

Given the location of the distortion relative to the visual carrier, CSO testing must be done with the modulation off, and CTB testing with the carrier off. For compliance measurements that is true. But it is inconvenient and disruptive to make such measurements across the cable channels and at each test point throughout the system. The spectrum analyzer allows you to look into frequency nooks and crannies that may give you a good indication of how your system is behaving without interrupting the flow of program material.

CTB beats lie right on top of the carriers. From the discussion of CTB frequency generation you know that the beats occur almost uniformly across the whole cable spectrum. To get a snapshot of CTB performance find an open spectral position above the last channel, or between channels 4 and 5, or below channel 2, and make the CTB tests using the procedures covered in the next sections. A record of performance at these frequency locations will help you to see performance trends over time, even though the test data may not be submitted for compliance. For CSO, whose strongest beats are offset from the carrier by 1.25 MHz, the area between channels is a good interstitial space to take your snapshot, especially where there is no lower adjacent channel, such as at channels 2 and 7.

Measuring CSO Without Turning Off the Modulation

Higher performance spectrum analyzers now offer a function known as time gating, which allows the analyzer to gather spectral information in small doses timed to coincide with one or more quiet vertical intervals of the television transmission. During this time the channel's modulation is "off" so that the distortion and interference impressed upon the carrier can be measured without actually turning off the channel's modulation. CTB cannot be measured with this noninterfering technique since the beat is right on the carrier, which is never "off."

Error-Free Measurements the First Time

This section formalizes the compliance test procedures and gives you some training to maximize your test efficiency and accuracy. The topics include time averaging of distortion measurements, correction for distortion near noise, distortion prevention in the analyzer, and detailed measurement procedures with examples.

Getting the Most from the Spectrum Analyzer

The CSO/CTB tests demonstrate the versatility and utility of the spectrum analyzer. With the spectrum analyzer you can look in the hidden corners for out-of-channel beats with just a few key strokes or knob turns. Simply set the resolution bandwidth to 30 kHz and the video filter to 30 Hz, and the distortion can be measured directly. That's all there is to it!

Set the analyzer for a resolution bandwidth of 30 kHz and a video bandwidth of 30 Hz, or, if your analyzer has digital video averaging, use that function with an average of 5 and set the video bandwidth to 1 kHz.

Composite Distortions over Wider Bands

An analyzer resolution bandwidth of 30 kHz is chosen because it is wide enough to gather the composite signals, but narrow enough not to take adjacent signal power into the amplitude reading. Because system carrier frequencies are often offset to prevent interference with aeronautical navigation signals, 30 kHz may not be sufficient to gather all the coherent disturbances under one composite reading. Figure 90 demonstrates this concept by measuring the CSO below channel 2, at 54.0 MHz, in three different bandwidths. Several distinct CSO beats are seen in the 3 kHz resolution bandwidth trace, the lowest trace. The middle trace, taken with a 30 kHz filter, collects these into one composite response which is higher than any one of the individual signals. This represents the composite beat level. The signal off to the left in the 30 kHz trace is a beat that is not included in the composite level. To remedy this, the resolution bandwidth is increased to 100 kHz, resulting in the top trace.

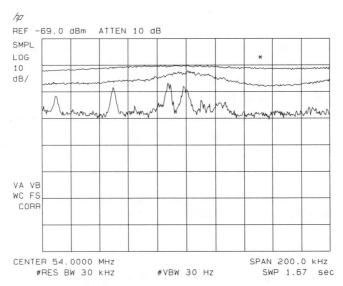

Figure 90. Measuring CSO with three resolution bandwidths. Top trace is with 100 kHz, the middle, 30 kHz, and bottom, 3 kHz.

CAUTION

A resolution bandwidth wider than 30 kHz may make a beat appear higher in level than actual because of the inclusion of adjacent channel modulation.

Unfortunately, the 100 kHz bandwidth is too wide, because it not only includes the signal on the left but also the vestigial sideband of channel 2. The solution would be to follow compliance guidelines and remove channel 2 modulation during the CSO measurement using a 100 kHz resolution bandwidth.

Composite Disturbances Change over Time

The nature of composite disturbances is that they are made up of so many other distortion signals that their amplitude can change over time. Often this change is not sufficiently averaged by the 30 Hz video filter, and measurement speed suffers because sweep time is slow. Digital video averaging is an alternative video filtering technique which provides longer averaging times while speeding the measurement.

The video filter is opened from 30 Hz to 1 kHz, enabling sweeps that are hundreds of times faster. The bulk of the averaging is now done by the analyzer's internal computer. The amplitude value from the video filter at each frequency point is averaged in with the resident average amplitude value according a built-in mathematical algorithm. You can set the number of averages taken by the analyzer. The lower the value, the greater the influence each new amplitude value will have on the overall average. For composite beat measurements, a video average of five sweeps is a good compromise between composite signal variation and efficient measurement speed. For more information on the operation and use of video averaging, see the Peterson reference listed at the end of this chapter.

TIP

Video filtering and video averaging automatically average the variation of beat amplitude over time.

The 1 kHz video bandwidth allows the use of wider frequency spans while maintaining a fast sweep speed. This means that you can measure all the CSO/CTB beats in one trace display. Figure 91 shows the CSO of Figure 90. The CSO is the small sharp signal at the center frequency.

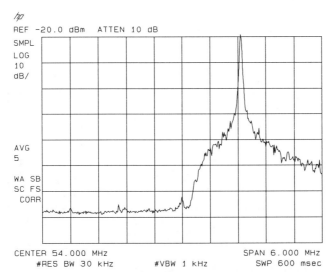

Figure 91. CSO observed with video bandwidth of 1 kHz and video averaging.

Correcting CSO/CTB for Analyzer Noise

When a signal is close to the internal noise of the analyzer, its value appears higher on the display than its actual value because the analyzer's detection circuitry cannot separate internal and system noise from input signals. When the signal is 10 dB or more out of the noise, the signal is so high above the noise that no correction is needed. Although the beats appear as CW signals, their composite nature makes them appear to the analyzer as noise. This means that the correction for beats is based upon the ratio of system noise to the amount of system noise being displayed, just as determined in the carrier-to-noise ratio measurements.

The procedure is simple. For beats that are within 10 dB of the displayed noise of the analyzer, perform the disconnect test. If this test shows a 10 dB or greater drop in noise, no correction is needed. If the difference is less than 10 dB then perform the disconnect test and apply a correction factor found in Figure 92.

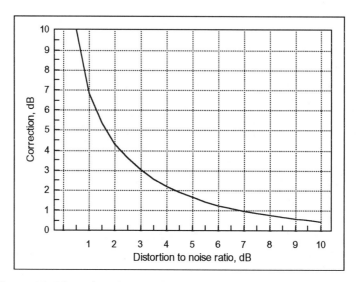

Figure 92. Distortion signal near noise correction graph.

Example 39. Correcting for signal-near-noise.

A CSO beat is measured at −36.3 dBmV. It is 3.6 dB higher than the adjacent displayed noise. From the C/N measurement procedure, you know that the noise drops 6.49 dB when the system signal is disconnected from the analyzer. What is the CSO amplitude?

Since the beat is less than 10 dB out of the noise, it is necessary to find the correction for the contribution of the analyzer noise. Since the noise drop was already determined when you measured C/N under these conditions, no additional testing is necessary. For a disconnect test value of 6.49 dB, a correction of 1 dB to the CSO level of 1 dB is drawn from Figure 92. The amplitude of the CSO is −36.3 dBmV − 1 dB = −37.3 dBmV.

Remember that when the beat is close to the noise and the system noise is less than 10 dB higher than the analyzer noise, the composite signal will always appear higher (worse) than it actually is. Therefore, if the beat passes the specification on the first inspection, there is no need to make the amplitude corrections.

When the beat is close to the displayed noise and the analyzer noise is within 10 dB of the system noise, the beat will appear at a higher level than actual. It's the worst case, so you may not wish to bother correcting its amplitude.

Preventing Overload

The spectrum analyzer has a wide band input mixer that can create distortion signals that look just like CSO/CTB. Chapter 2 discussed the analyzer's susceptibility to overload in terms of signal compression. Signal levels that cause the creation of distortion products are much lower in power than those that cause compression. Fortunately, the test and cure for internal analyzer distortion is just as easy as for compression. Simply decrease the power to the analyzer's mixer by increasing attenuation in the signal path, and view the change of the distortion products. If the distortion changes as much as the attenuation changes, then the analyzer is not distorting the input. Increase the attenuation until the distortion signal levels agree with the attenuation steps.

Since the beats are small, use 2 or 3 dB attenuation to determine if the distortion is coming from the system or the analyzer. If the analyzer is without overload, the beat should only change level by the amount of attenuation inserted.

The analyzer's internal attenuator step is usually 10 dB. Since most composite beats are only 5 to 15 dB out of the displayed noise of the analyzer, an increase of attenuation by 10 dB will bury the distortion signal, preventing an accurate look at the signal step size. The solution to this is to use an external step attenuator which can introduce smaller steps. Steps of 2 or 3 dB are recommended. Although it is usually not necessary to use a preamplifier for CSO/CTB tests, if one is in place, make sure you place the step attenuator ahead of the preamplifier to test both the analyzer and preamplifier for overload.

If a preamplifier is used, test both analyzer and preamplifier for overload, not just the analyzer.

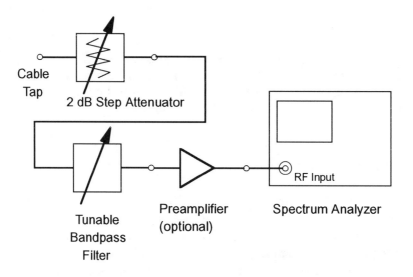

Figure 93. Setup for prevention of analyzer distortion.

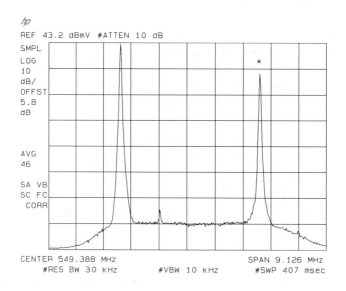

Figure 94. CSO measurement using a bandpass filter.

If the addition of sufficient attenuation to prevent internal distortion also makes CSO/CTB measurements difficult by putting the beats too close to the displayed noise, a bandpass filter must be inserted in to the signal path. These combinations are shown in Figure 93.

The bandpass filter must have a pass band wide enough to view the distortion. Figure 94 shows a CSO measurement using a bandpass filter. The filter's frequency response is noted by the roll-off of system and analyzer noise on both sides of the channel. It is not necessary to use a single channel bandpass filter. Often the power to the analyzer can be sufficiently reduced with a multichannel filter.

Measurement Procedures

Now let's apply all the lessons learned to make CSO/CTB measurements. Before starting, you need to decide whether you are making the measurement for compliance or for a system performance check. If compliance testing is required, you will have to have a channel or channels available that can be unmodulated and turned off entirely when required by the person at the subscriber tap.

Measure the Carrier

Measure the carrier level just as taught in Chapter 4, with one important condition: Try to leave the carrier peak at the reference level of the analyzer while measuring the beats. This insures that CSO/CTB is measured as a relative measurement with the best accuracy possible.

The choice of carrier will depend upon your testing requirements. If you are looking for system performance characteristics and not compliance testing, then a carrier near the area where CTB or CSO is to be sampled will do. Otherwise the carrier to be measured must be the carrier to be unmodulated and turned off. Once again, remember that turning off the carrier modulation and/or carrier should be done without turning off the active devices in its transmission path. Otherwise the distortion will disappear with the carrier, and you won't get a true beat reading.

| TIP | **For quick and accurate measurements, leave the carrier level at the analyzer's reference level while measuring the nearby beats.** |

Out-of-Channel CSO Measurement

When measuring CSO for monitoring your system's performance, there is no need to turn modulation off. But you do need to look in the system nooks and crannies where adjacent channels do not infringe upon the distortion's most likely frequency spot. One such area is often just below channel 2. Here is the procedure for measuring CSO near a modulated channel.

Example 40. Measure CSO of a modulated channel.

Channel 2 may not have a lower adjacent signal. Measure the CSO at that point.

Set the spectrum analyzer to center channel 2 carrier, at 55.25 MHz with the span at 6 MHz, and resolution and video bandwidths of 300 kHz. Set the trace mode to maximum hold and adjust the reference level until the signal peak is at the reference level. If you overshoot the top graticule, it is necessary to change the trace mode back to its normal mode. Raise the reference level to get the carrier peak below the reference level, then repeat the maximum hold process. The display looks like Figure 95.

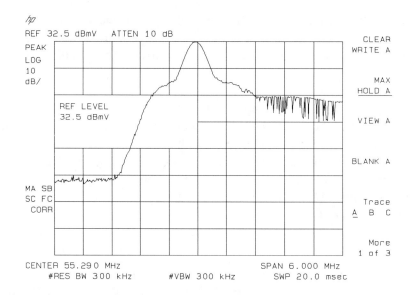

Figure 95. Measurement of the carrier.

If the reference level control does not have the resolution necessary to let you set the peak signal exactly on the reference level, use this procedure: Move the signal below the reference level and begin the maximum hold trace buildup again. Place a marker at the peak of the signal after it has a uniform peak, and then use the marker-to-reference-level function to bring the signal to the top graticule.

From this point on be careful not to change the reference level, which now represents the carrier level. Put the trace in its normal peak mode and select the sample detection mode. Change the resolution bandwidth to 30 kHz, the video bandwidth to 1 kHz, and turn on video averaging with a sample of 5. If your analyzer does not have digital video filtering, set the video filter to 30 Hz.

The distortion, if any, will appear as a signal response 1.25 MHz below the carrier, or at 54.0 MHz for channel 2. If this signal and/or the noise is within the lowest display division of the analyzer, and your analyzer is not calibrated for that division, decrease the reference level by 10 dB. Place a marker on the CSO peak, and note the amplitude. Figure 96 shows the a beat of −35.0 dBmV.

The carrier to CSO, sometimes written C/CSO, is the reference level, corrected for any changes, minus the CSO amplitude. In this example the reference level was increased by one 10 dB step. So CSO = +22.5 dBmV + 10 dB −(− 35.0 dBmV) = 67.5 dB.

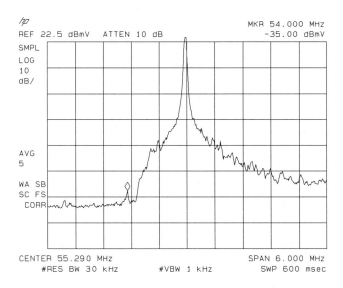

Figure 96. CSO signal below channel 2.

Because the signal is close to the noise, it appears higher than actual; that is, the CSO is actually better than 67.5 dB. Since the beat is within 10 dB of the noise, it is likely to appear higher than actual. The correction value depends upon the disconnect test. For a disconnect noise drop of 6.5 dB, the correction value from Figure 92, is 1 dB. The CSO is 67.5 dB + 1 dB, or 68.5 dB. While it is unlikely you will want to make signal-near-noise corrections to such an excellent CSO reading, the example is given to show the correct process.

It is possible to view CSO between channels, although the aural modulation of the lower channel may mask very low CSO values. In order to get a quick scan through a group of channels, set the analyzer's controls to what they were at the end of the last example. Then with the center frequency step size set to 6 MHz, step between channels until you see larger CSO distortion beats. Measure the higher levels. The system slope or frequency response may have significantly changed the carrier's level, so if you find an extraordinarily high beat, remeasure its associated carrier level. Figure 97 shows such a between-channel beat. Remember that the measurement of CSO with modulation means that you are measuring the beats outside the video modulation range of the channel, that is,

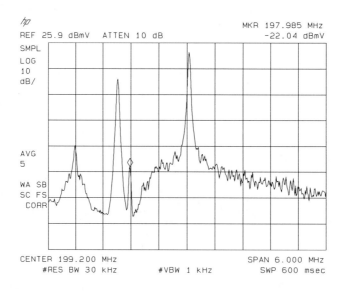

Figure 97. A between-channel CSO beat.

the −1.25 MHz beat. The +1.25 MHz beat is the one that interferes with program material, and must be measured for FCC compliance tests with the modulation off.

TIP

The −1.25 CSO beat can be measured without interrupting subscriber service, and is a good indicator of your system's distortion performance.

For efficiency's sake, CSO compliance testing is done along with the CTB test, covered in the next section.

In- and out-of-Channel CTB Measurements

Measuring CTB without turning off the channel carrier is more difficult than with the CSO measurement because there are fewer out-of-channel CTBs. Informal CTB testing requires that you have an unoccupied portion of spectrum amidst the channels, where CTB is often at its worst. One such a spot is the space between channels 4 and 5. The CTB here is not under a carrier because of the channel 5 frequency offset. Unfortunately, this space is often used for intercable services, such as digital audio signals as shown in Figure 98.

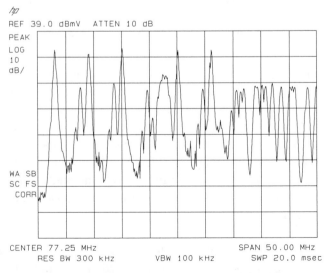

Figure 98. This space between channels 4 and 5 in this system is not available for CTB testing.

Another spot to check system CTB performance is above the last channel. Set the analyzer to step in 6 MHz increments; center the highest system channel; and step the center frequency up until an unoccupied area is reached. Set the analyzer to search for composite beats and narrow the span until you see some signal responses. Figure 99 shows the results of such a search.

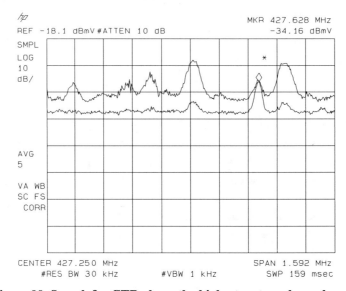

Figure 99. Search for CTB above the highest system channel.

The span in the figure is narrowed to 1.6 MHz to detect the signals that make up CTB in this region. Test for overload. The upper trace shows the response with 0 dB of input attenuation, and the lower trace is with 10 dB of input attenuation. The difference in the amplitudes in all but the marked signal means that the analyzer with 0 dB attenuation is overloaded for these composite triple beats. Since the signals are too close to the noise to change the attenuator to its 20 dB setting, an external 2 dB attenuator must be used. When the analyzer's distortion is eliminated by inserting the right amount of attenuation, the CTB level measurement is made just as for CSO in the last example, using the closest carrier's level as a C/CTB reference.

In Figure 100 both CTB and CSO can be observed in the same channel spacing. The left marker points to the carrier frequency slot, now occupied by system CTB, and the right marker shows the +1.25 MHz CSO. The CSO is a cluster of five signals with a spacing that probably is indicative of the cumulative frequency variances throughout this standard frequency allocation system. The highest CSO is right at +1.25 MHz, however. Note that the amplitude scale has been changed to 5 dB per division for clarity of the figure. Changing the scale should not be done except to investigate the nature of the beats more carefully because such a change adds considerable uncertainty to the amplitude accuracy of the measurement.

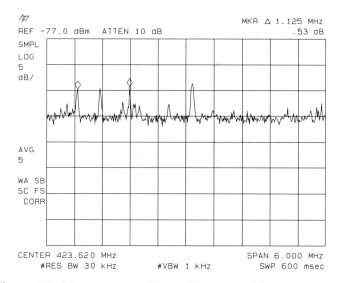

Figure 100. Measurement of both CTB and CSO where no channel roams.

Compliance CSO/CTB Testing

Because the compliance testing of CSO/CTB requires the removal of modulation and carrier of a channel, it is good practice to measure these when you measure other interfering measurements, such as C/N and in-channel frequency response. Here is an example of the procedure.

Example 41. Measure CSO/CTB for FCC compliance testing.

Make a CSO/CTB compliance test on a single channel.

Pick a channel whose modulation and carrier can be turned off and on at your command. In a 6 MHz span, measure the amplitude of the modulated carrier with the resolution and video bandwidths set to 300 kHz. Set the reference level equal to the carrier level with the reference level control. Turn the detection mode to sample, and then switch the trace out of maximum hold mode. With resolution bandwidth set to 30 kHz, video bandwidth set to 1 kHz, and video averaging set to 5 sweeps, have the channel's modulation turned off. Look to the upper frequency side of the carrier for CSO beats at 0.75 and 1.25 MHz away from the carrier.

In Figure 101, the CSO products are clearly visible with the carrier and modulation off. CSO is not a product of the individual channel's processing, but rather of the amplification chain in distribution.

Test the beat levels for analyzer distortion caused by overload by inserting 2 dB between the analyzer and the cable. The distortion signals should change by the same 2 dB. If they change more, additional attenuation must be added to the signal. If this attenuation puts the distortion beats below the displayed noise, then reduce the power to the analyzer by using a channel or channel bandpass filter. Repeat testing of the analyzer for overload after either of these measures are taken.

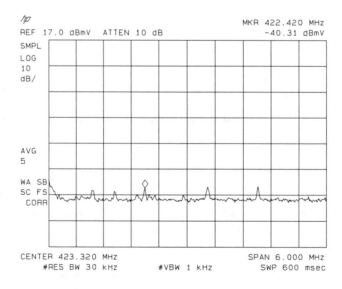

Figure 101. FCC testing for CSO/CTB.

If the CSO and/or displayed noise are in the lowest division of the analyzer, lower the reference level by 10 dB to move the signal into the next highest division. Place a marker at the peak signal of either CSO signal cluster, let the analyzer sweep at least five more times, and record the marker amplitude. The amplitude is −40.3 dBmV for the +1.25 MHz CSO.

Since the disconnect test shows a noise difference of 6.5 dB, the CSO is actually 1 dB lower, using the procedure in Example 39. The CSO amplitude is −40.3 dBmV − 1 dB = −41.3 dBmV.

Compute the C/CSO: the carrier level +17.0 dBmV −(−41.3) = 58.3 dB, considerably better than the FCC recommendations.

The CTB is measured at the carrier frequency, 421.25 MHz in this case. Make sure that the carrier has been removed without powering down amplifiers in the channel path. See Figure 102.

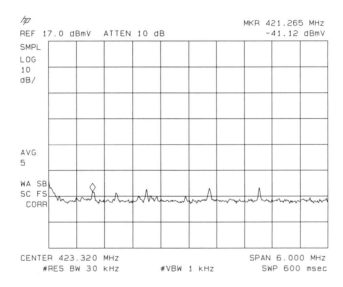

Figure 102. CTB measurement at a channel with the carrier turned off.

The CTB level is measured and corrected as in the CSO example above. In the figure the CTB is −41.1 dBmV. The same noise correction of 1 dB is applied, so the C/CTB is +17.0 dBmV −(−41.1 − 1) = 59.1 dB.

CAUTION

When removing the carrier for a CTB test, don't break the signal path between channel sources and the tap, or your beat readings will not show the true system performance.

How Accurate Are these Measurements?

Most of the elements causing uncertainty have been minimized in the procedures described above. Setting and leaving the carrier level equal to reference level minimize the reference level uncertainty. The scale fidelity uncertainty cannot be eliminated because of the wide amplitude range of the measurement. Frequency response of the analyzer will only be a concern when measuring CSO/CTB against a carrier that is more than one channel away from the beats, such as when the beats are measured between channels 4 and 5, or above the highest channel.

Example 42. Evaluate the accuracy of Example 41.

In Example 41, the C/CSO was measured at 58.3 dB, including the correction for signal near noise. This distortion value is relative to the analyzer's reference level, and therefore needs to include only scale fidelity. Looking in the analyzer's performance specifications for relative amplitude, Chapter 4, Table 2, under "Display Scale Fidelity." The log max cumulative is ±0.75 dB for a range to −60 dB. So, including the uncertainty, the CSO is a value between 59.05 and 57.55 dB.

If the reference level had been changed to bring the CSO and noise out of the lowest display division, the CSO would also include the reference level uncertainty. The specification for this uncertainty is about ±0.3 dB. Under this condition, the total uncertainty is ±0.75 dB ± 0.3 dB = ±1.05 dB. The CSO is a value between 59.35 to 57.25 dB.

Other errors are more subtle. The spectrum analyzer itself may have different distortion measurement capability across its frequency range. This means that you should always test the analyzer for overload when making a measurement at a level or frequency not repeated often nor made with the same spectrum analyzer. Another potential error is the frequency response of the bandpass filter used to reduce the overload of the analyzer. Tunable bandpass filters are trouble because it is their nature to change frequency response as they are tuned. With this type filter it is necessary to maximize each response measured at each step. A typical band pass filter response, as measured with a spectrum analyzer and tracking generator is shown in Figure 103. The amplitude variation across the filter's pass band is as much as 3 dB, potentially adding error to the CTB and CSO measurement. Wider, fixed-tuned bandpass filters usually offer better frequency response and are easier to use because they do not have to be tuned for each channel change. Filter frequency response is characterized by either data from the manufacturer or by testing the filter with a scalar analyzer or combination spectrum analyzer and tracking generator.

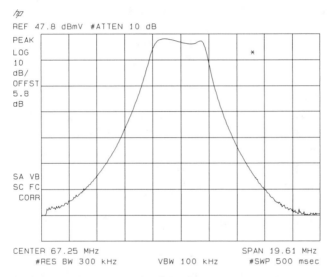

Figure 103. Typical bandpass filter frequency response.

TIP **Usually the CSO/CTB measurement is accurate to about 1.5 dB.**

Summary of Procedures

CSO/CTB distortion is the most troublesome distortion product in cable systems with more than 20 channels.

* With standard frequency allocation systems CSO appears ±0.75 and ±1.25 MHz from the visual carrier. CTB is found at the carrier frequency.
* Composite beat distortion signals have noiselike signal characteristics that require amplitude averaging for realistic results.
* FCC regulations require CSO testing in the video portion of the channel. This means turning off the channel modulation. However, merit testing can be done by using the spectrum analyzer to measure the lower 1.25 MHz beat without interrupting programming.
* CTB compliance testing requires that the channel carrier be turned off, although system merit testing can be done at frequencies where channels are absent.

The spectrum analyzer settings for optimal CSO/CTB measurement speed and accuracy are as follows:

* Measure the carrier level first using 300 kHz resolution and video bandwidths by placing the carrier peak at the reference level.
* For most measurements, use 30 kHz resolution bandwidth, 1 kHz video bandwidth, and digital video averaging of five averages.
* For analyzers without digital video averaging capability, use 30 kHz resolution bandwidth and 30 Hz video bandwidth.
* Use average or sample IF detection.
* Test for overload with 2 or 3 dB external attenuator when beat is close to the noise floor.
* If distortion within 10 dB of the displayed noise floor reads too high, it may be corrected using the disconnect test and the noise-near-noise correction chart.
* Display scale fidelity and reference level accuracy account for the most of the analyzer measurement uncertainty. Bandpass filter response can also cause frequency response errors.

Selected Bibliography

* Adams, Mark. *Composite Second Order: Fact Or Fantasy*. Technical Papers of NCTA, pp. 233-237, 1988.
* *CED: Technical Standards Supplement*. Communications Engineering & Design magazine, Chilton Publications, Denver, CO, July 1993.
* Edgington, Francis M. Hewlett-Packard Company, Microwave Instruments Division, Santa Rosa, CA, *Personal communication*, August 1994.

- Goehler, Randy. Cox Cable of San Diego, CA, *Personal communication*, May 1994.
- Peterson, Blake. *Spectrum Analysis Basics*. Hewlett-Packard Company, Application Note AN 150, Literature No. 5952-0292, Santa Rosa, CA, 1989.
- Pike, Dan. *The Measure and Perceptibility of Composite Triple Beat*, Technical Papers, 28th Annual NCTA Convention, National Cable Television Assn., 918 Sixteenth N.W. Washington, DC, 1979.

Cross Modulation
Investigated

9

Overview

Cross modulation is the distortion most troublesome for cable systems with fewer than 20 channels. Distribution amplifier technology has changed along with the increasing number of channels systems must carry. With these changes the nature and measurement of cross modulation has changed too. This chapter looks at the makeup of cross modulation and offers some guidance in the sometimes controversial ways in which it is measured in systems of all sizes. The spectrum analyzer is used as a frequency domain analyzer and as a fast Fourier analyzer to make these measurements.

The Measurement Confusion

It is not a mistake that neither the FCC compliance rules and regulations nor current NCTA recommended practices have cross modulation listed as separate distortion measurements. Cross modulation is another form of third-order distortion caused by the amplification of groups of channels by distribution amplifiers. You will find abundant references to cross modulation in the test and system design specifications for distribution amplifiers and system design procedures, however. This is because cross modulation is more a distribution amplifier specification than a well-defined system compliance spec. That is not to say it is not important! Cross modulation can bring out the irate calls and truck rolls with the best, or worst, of the subscriber compaints.

Figure 104. Cross modulation television receiver simulation.

Why Cross Modulation Causes Truck Rolls

Subscriber complaints usually cite a channel interference that looks like wiper blades on the picture, sometimes chopped up into segments, sometimes as a matrix pattern floating over the picture. Figure 104 is a photo simulation that illustrates the latter of these effects.

The Measurement in Brief

There are two ways to measure cross modulation with the spectrum analyzer: in the frequency domain, and with a special function called fast Fourier transform. Here are the steps for the frequency domain measurement.

1. Select a channel and turn carrier modulation off.
2. Center the unmodulated carrier in a span of 50 kHz, and measure and record its amplitude in dBmV.
3. Set the resolution bandwidth to 1 kHz, and the video bandwidth to 300 Hz.
4. Change the display mode to maximum hold and let the analyzer sweep for about 30 seconds or until the 15.75 kHz sidebands have grown as much as they can.
5. Measure the amplitude of the sideband.
6. Calculate the cross modulation in decibels by subtracting the sideband amplitude from the carrier amplitude.
7. If your analyzer has the fast Fourier transform function (abbreviated FFT), make the measurement again, and compare readings. The FFT cross modulation reading should be lower.

Cross modulation is less common in larger systems simply because it is overrun by the other major third-order distortion, CTB. However, certain channels in a large system can be particularly susceptible to cross modulation, so it is well to understand how it looks on both television receiver and spectrum analyzer in order to help solve customer complaints.

FCC Regulations

Cross modulation is governed under the distortion portion of the compliance specifications, FCC Rules and Regulations Performance Tests in part 76.601 (c) (8) and Measurements procedures 76.609 (f). See the Chapter 8 for the full text of these regulations.

What Distortion Regulations Mean for Cross Modulation

♦ Cross modulation must be below 51 dB for systems where the carriers are not tied to a single synchronizing generator.

- Cross modulation must be below 47 dB for systems where the carriers are tied to a single synchronizing generator.
- A spectrum analyzer can be used to make the measurement.
- Measurements must be made with no modulation on the carrier.

But, as you will see, there are enough concerns about how cross modulation is perceived by the subscriber, the effects of program material on its observation, and the way it is measured, so that no strict specification can be pointed to with confidence. Add to this the fact that a majority of cable systems today have 40+ channels, where CTB/CSOs are the dominant distortion, and you will see the dilemma.

TIP

Measure cross modulation to the same distortion levels as you do for CSO/CTB.

Origins and Definition

Like carrier-to-noise and CTB/CSO, cross modulation originates in the cable amplifiers. It can't be said that the distortion is generated, because one definition of cross modulation says that it is a third-order distortion which does not result in a new beat. Cross modulation is just what the name implies; it is a third-order effect that modulates the modulation within the distribution amplifiers themselves. It produces amplitude modulation on the dominate system signals, the visual carriers.

Since most of the modulation energy of a cable system is concentrated in the horizontal sync pulses, the 15.75 kHz sideband is the one that is modulated onto unsuspecting carriers. Figure 105 shows just how much sideband energy of the sync pulses is in the spectrum of a normally modulated carrier. The distortion sidebands are masked by the carrier's normal modulation, and, therefore, cannot be observed while the channel is modulated. Broadcast television standards require that the sync tip of the carrier be held at a constant level. Cross modulation, being an unsynchronized amplitude modulation, changes the levels of the sync tips. Figure 106 shows this effect.

Figure 106(a) represents the normally modulated video carrier, greatly simplified, of course. The effect upon an unmodulated carrier is shown in (b). On the left is the unmodulated carrier and on the right the waveform with amplitude modulation in the form of the sync pulses and black bar impressed on its envelope. The amplitude modulation is

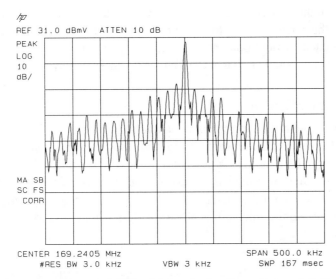

Figure 105. The 15.75 kHz sidebands dominate the modulation of the visual carrier.

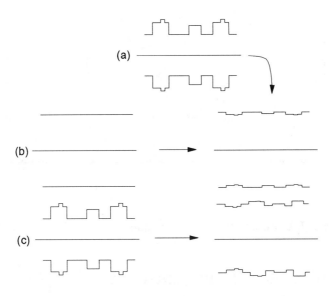

Figure 106. Effect of a video carrier (a) amplitude modulating an unmodulated carrier (b) and another modulated carrier (c).

shown as down from the carrier peak swing, but this is arbitrary. The modulation could just as well add to the carrier voltage to swing beyond the carrier's nominal level. In (c) the effect of one modulated carrier upon another is shown. The sync tips of the transmitted carrier are modulated by another channel's modulation, indicated by the dotted envelope.

Cross Modulation Definition

The way this effect is quantified for comparison and specification is shown in Figure 107. The total amplitude swing of the unwanted amplitude modulation is compared to the peak voltage. This value can be written as a percent, but most of the time it is given a value in dB. Since the signals are represented as voltage swings, cross modulation in dB is 20 log(a/b). This is a convenient ratio mathematically because the voltage swing cannot exceed the peak swing, so the value a/b is always be less than 1, and the cross modulation in dB is a negative value.

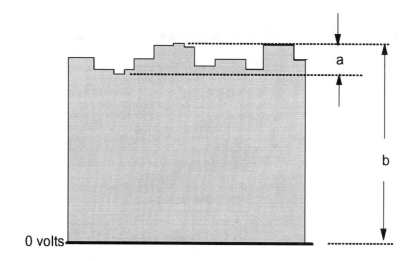

Figure 107. Definition of the cross modulation level.

Cross Modulation Not without Controversy

On the face of it, the cross modulation measurement should be a simple one. Conceptually, all that has to be measured is the remaining 15.75 kHz sideband on a carrier whose modulation has been turned off. But three areas of concern open the cross modulation measurement methods and specification levels to question. These concerns are:

- ◆ Past NCTA-recommended measurement practices disagree with some spectrum analyzer test procedures by their manufacturers.
- ◆ Current distribution amplifiers produce cross modulation products rich in phase information, which may cause a spectrum analyzer to read higher distortion than actual when it is measuring in frequency domain mode.
- ◆ Cross modulation perception levels vary widely depending upon test and programming materials.

Let's look at each of these before making procedure recommendations.

Past Test Procedures

Cross modulation has been measured various ways, depending upon the situation and availability of test equipment. Figure 108 illustrates these by showing the measurement parameters in the time domain. The spectrum analyzer commonly makes the measurement (a) as a ratio of the unmodulated carrier to the cross modulation sideband. A past NCTA-recommended practice was to use the ratio of the 50% 15.75 kHz sideband to the cross modulation level, shown as ratio (b).

It is obvious that following the two measurement definitions gives different cross modulation results; the spectrum analyzer adds as much as 10 dB. To conform to the definition of cross modulation in (b), 10 dB is subtracted from the spectrum analyzer reading.

Cross Modulation Is AM and PM

Twenty years ago cross modulation was thought to be caused by a simple mechanism in the distribution amplifiers; the gain of the amplifiers was modulated by the host of signals amplified. And this is true for the channels in the lower frequencies; cross modulation is caused by the AM within the amplifiers. At the high-frequency channels, with changes in the technology of amplifiers, cross modulation takes the form of phase modulation, PM. This phase modulation causes minute and temporary shifts in the carrier frequency, which in turn causes a slope detection amplitude change in the pass band of the channel. In other words, the PM results in an AM.

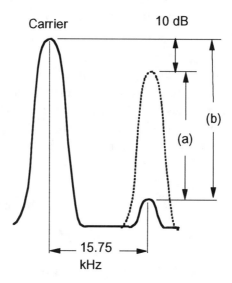

Figure 108. Cross modulation measurement ratios for (a) NCTA-recommended practices and (b) the traditional spectrum analyzer method.

The spectrum analyzer cannot measure the phase summation of signals. Its peak detection scheme takes the highest value of the signals in the pass band of the resolution filter regardless of their phase relationships. This means that the spectrum analyzer may display cross modulation due to PM higher than a wave analyzer would. However, one new measurement technique available in some spectrum analyzers is a method called fast Fourier transform, or FFT. This measurement technique offers some help in correcting for phase errors. More details and examples are found in the measurement section later in this chapter.

Seeing Cross Modulation on the Television Receiver

Levels of cross modulation that can be perceived vary widely because of the subjective nature of the tests and the selection of programming material used. Cross modulation seen against a predominantly white luminance with still subject matter is easier seen than against dark and busy pictures. Values of cross modulation perception levels from 43 to 63 dB are mentioned in the literature.

These considerations place the FCC-recommended distortion level of 51 dB for nonsynchronized carriers in the middle, a good place to start when the corrections for past practices and implications of phase modulation are considered.

Cross Modulation Measurement in the Frequency Domain

The most straight forward way to measure cross modulation with a spectrum analyzer is to measure the 15.75 kHz sideband of an unmodulated carrier. Before making this measurement, follow the guidelines for preventing overload of the spectrum analyzer described in Chapter 8, page 177.

Example 43. Measure the cross modulation in the frequency domain.

Select a channel whose modulation can be turned off temporarily for cross modulation tests. Be careful not to disconnect active components between the signal source and test point, or the cross modulation test may be invalid.

Set the analyzer center frequency to the channel carrier, and set a span of 50 kHz, with a resolution bandwidth of 1 kHz and a video bandwidth of 300 Hz. Set the peak carrier at the reference level. The value in this example is +29.3 dBmV. The sidebands of analyzers with 80 dB display ranges often be in the lowest division, which may be uncalibrated. Raise the signal level on the display by increasing the reference level 10 dB. Place the analyzer display in maximum hold and wait about 1 minute while the sidebands fill in. Figure 109 shows this display.

Place a marker on the highest part of the 15 kHz sideband. The level is −34.9 dBmV. The cross modulation is the difference between the carrier level and the sideband level, +29.3 −(−34.9) = 64.2 dB.

In this measurement, the unmodulated carrier is compared to the sideband directly. Even when a 10 dB cushion is applied to this cross modulation, the distortion is better than the FCC-recommended 51 dB by about 3 dB. When higher distortion levels are measured that trigger no subscriber complaints, accept the distortion level as a test point reference for future measurements.

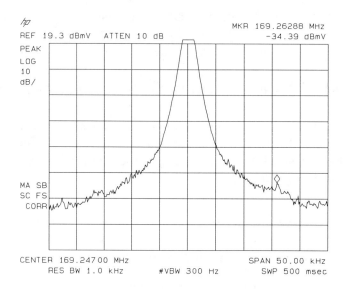

Figure 109. Cross modulation measurement in frequency domain.

Consistent cross modulation levels measured at the same test point, coupled with a lack of subscriber complaints, may be your best measurement technique.

Using FFT to Measure Cross Modulation

The fast Fourier transform function, built into some spectrum analyzers, operates like a second spectrum analyzer within the analyzer. FFT is an internal computer process that takes a gulp of carrier voltage amplitude data for a specified amount of time, and transforms it into a spectral response of the video spectrum. The analyzer is tuned to the center of the carrier, and the span of the analyzer is set to zero; that is, the analyzer is "fixed-tuned" to the carrier. This concept illustrated in Figure 110, which shows the analyzer tuned to the carrier under test. The analyzer's heterodyne oscillator is stopped from sweeping so that only one input signal frequency is being passed to the analyzer's IF. From the front panel of the analyzer, this is known as the zero span mode.

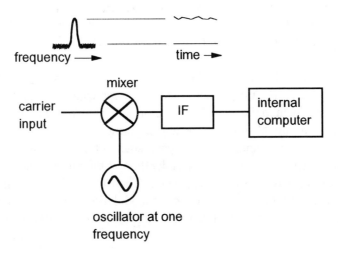

Figure 110. Spectrum analyzer functions as a fixed tuned receiver for using its FFT function.

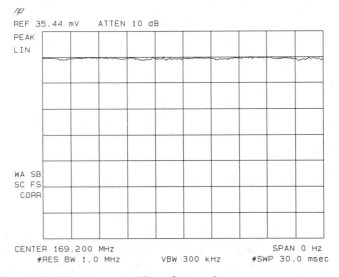

Figure 111. Zero span reading of a carrier.

Now the display shows the carrier amplitude over time, not frequency. Figure 111 shows an example. The analyzer amplitude has been set to linear mode to more clearly see amplitude variations of the signal. The resolution bandwidth is set to 1 MHz in order to allow the carrier's modulation into the IF for FFT processing.

The FFT function stores the time-domain data and performs a mathematical transformation to give its Fourier, or harmonic signal contents. The result is a graphical representation of the "demodulated" video waveform, as shown in Figure 112. The results of this test on the same signal shown in the above example demonstrates the difference between the measurement techniques. The measurement of cross modulation in the time domain is a peak measurement, with no consideration of the phase relationships between the constituents of the cross modulation products. The FFT method, which does reconstruct the phase relationships of the carrier modulation, reads considerably less cross modulation level. In fact, it is not visible in the display, a value better than 70 dB.

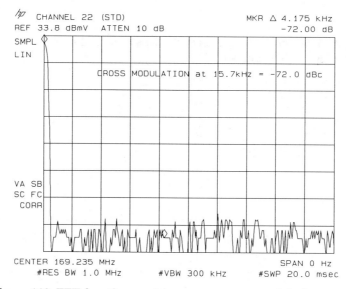

Figure 112. FFT function used to measure cross modulation.

Note that the FFT marker is not placed at the 15 kHz frequency, but at about 4 kHz. This is because this particular FFT function produces subharmonics of the amplitude modulation, called aliasing responses, which represent the fundamental sidebands. Although the theory and operation of the FFT function goes beyond the scope of this book, its use can be found in your spectrum analyzer's operation manual and supporting application notes.

TIP

FFT function of the analyzer takes phase content of cross modulation distortion into account.

Accuracy Considerations

Accuracy depends upon knowing the exact specification. As discussed, there are a number of ambiguities in the interpretation and measurement of cross modulation. The most important advice is to avoid overload conditions, and to make your measurements as consistent as possible.

Conclusions

The spectrum analyzer makes very repeatable measurements, so once you have established cross modulation levels for different channels at different test points, use the analyzer to compare results over time. This is your best guarantee that your system is meeting its cross modulation specifications. For systems with more than 40 channels, CTB/CSO should take the majority of your distortion test time, for these are the dominant trouble makers.

Summary

- Pick a cross modulation measurement technique and stick with it.
- Set your own specification levels based upon test technique and subscriber satisfaction. Don't worry about absolute distortion level so much as changes to the levels over time.
- Even if no subscriber complaints are registered, measure cross modulation periodically to look for system trends.

Selected Bibliography

- Adams, Mark and Pidgeon, Rezin. *Cross Modulation: Its Specification and Significance*. Communications Technology Publications Corp., April 1987.

- *Cable Television System Measurement Handbook.* Hewlett-Packard Company, Literature No. 5952-9228, Santa Rosa, CA, January 1977.
- Langenberg, Earl. *Cross-Modulation Redefined.* Communications Technology Publications Corp., November 1990.
- Luettgenau, G. G. *Cross-Modulation in HRC Systems.* paper from TRW Semiconductors, Lawndale, CA.
- Simons, Ken. *Technical Handbook for CATV Systems.* Jerrold Division, General Instrument Corporation, Publication No. 346-001-01, 3rd ed.

10

Modulation Distortion at Power Frequencies–Hum

Overview

This chapter will help you understand and measure low-frequency disturbances, specifically those related to power-line frequencies. The term hum is used to lump these together for convenience. Measuring hum is one of the easiest tests to perform, and doing it often, in different locations throughout your plant, is one of the basic measurements in preventive maintenance. Excessive hum is usually an indicator of power and connection trouble, and perhaps pending hardware failure. The spectrum analyzer can easily help you see small changes in hum levels long before they cause customer complaints.

Hum Annoyance Factor

Hum, especially power-line related hum, causes rolling bars on the television picture at about the 2% level, depending upon the sensitivity of the television receiver. Figure 113 shows this effect simulated on a photograph. At 4% the power line bars begin to tear the picture with one or two bars on screen at a time.

Figure 113. Simulation of 3% power line hum on the television receiver.

FCC Regulations

Technical Standards part 76.605 (a) (10): The peak-to-peak variation in visual signal level caused by undesired low frequency disturbances (hum or repetitive transients) generated within the system, or by inadequate low frequency response, shall not exceed 3 percent of the visual signal level. Measurements made on a single channel using a single unmodulated carrier may be used to demonstrate compliance with this parameter at each test location.

The Measurement in Brief

1. Tune and center to a modulated carrier.
2. Set the analyzer for resolution bandwidth of 1 MHz, video bandwidth of 1 MHz, sweep time of 30 ms, and linear display with voltage units.
3. Sweep the analyzer in the single sweep mode.
4. Find the maximum point with the marker.
5. Use the marker delta and find the minimum.
6. Percent (%) hum is (1.0 minus delta) value times 100.
7. Take the average of at least five readings.
8. If this average value is over 3%, measure again on an unmodulated carrier. The hum may be caused by program material, not the plant.
9. With modulation off, set the video bandwidth to 1 kHz, sweep time to 50 ms and measure as in steps 1 to 7.
10. The hum level is accurate to about 15% of the value in step 9.

What This Means

- 3% is the maximum level allowed, and is one of the most objectionable distortions to viewers.
- The distortion typically affects the whole transmitted bandwidth if it is found in one channel.
- Testing at each location ensures that the plant along different branches is healthy.

It is important to record levels and watch for trends of hum interference. Seeing levels move up to between 1 and 1.5% usually indicates something is going wrong. Monitoring hum levels makes good preventive maintenance sense because the failures of power supplies, connections, and errant amplifiers can take out an entire branch if not tended to early.

TIP

Excessive or changing hum levels are usually an indicator of pending hardware trouble.

Nature of Hum

Hum, by definition, is an amplitude modulation of the carriers on the cable. Amplitude modulation is the modulation used by the visual carrier to transmit the television receiver picture information, so hum appears on the picture, as shown in Figure 113. Hum amplitude modulation, or AM, is defined differently from what you may have studied. It is important to know this when using the spectrum analyzer to make the measurement.

> **CAUTION**   **The definition of hum amplitude modulation gives a value twice as large as does the definition for "conventional" AM.**

NCTA Definition

The NCTA hum definition is two times as much as the IEEE definition. There is no need to detail the differences. In fact, it is confusing to show both definitions, so only the cable television hum modulation is defined. In other words, the AM caused by low-frequency disturbances is the ratio of the peak-to-peak variation of the signal amplitude over time to the peak variation. These values are most conveniently expressed in volts, and the ratio as a percent. Figure 114 shows this definition and equation.

Note the similarity of hum definition to that of cross modulation. One difference is that hum is expressed in percent (%), while cross modulation is as a dB ratio.

How Hum Is Generated

The AM frequencies of hum are most often power-line related, that is 60 Hz and the second harmonic, 120 Hz, but low-frequency disturbances up to 400 Hz fall under the definition of low-frequency disturbances. At the low frequency side, the hunting of an amplifier's AGC, or the periodic hammering on a tower by wind-buffeted hardware can cause periodic modulation that is lower than 60 Hz. These frequency extremes are rare.

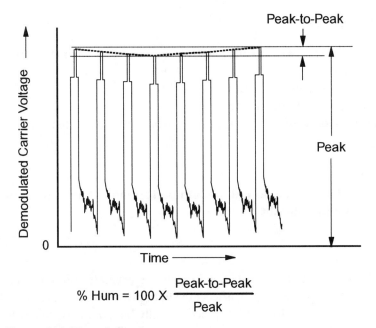

$$\% \text{ Hum} = 100 \text{ X } \frac{\text{Peak-to-Peak}}{\text{Peak}}$$

Figure 114. Hum defined.

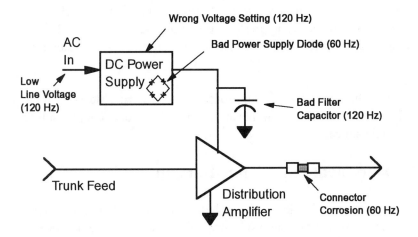

Figure 115. Common sources of power-line frequency hum interference.

More common are these causes:

- ◆ Low line voltage or wrong voltage setting on a distribution amplifier DC power supply (120 Hz).
- ◆ Failing filter capacitor (120 Hz).
- ◆ Bad power supply diode (60 Hz).
- ◆ Loose connector forming a diode junction from corrosion at the junction (60 Hz).

TIP   **When hum is found, tracing performance back up the branch should find the source.**

Hum is usually confined to the distribution branch it is generated on. For example, a bad diode on a feeder cable power supply affects the hum just on that branch, and then all the channels on that branch are distorted. But occasionally hum does not conform to this rule. If excessive hum is detected on only one channel, one of two phenomena are occurring. First, suspect the source, that is, the head end. The program material may already be contaminated before it gets to the head end, or by some rare occurrence, hum is added at the head end rack. Second, suspect a form of cross modulation in which one channel has the vertical interval sync pulses impressed upon it by another channel. This type of hum is unusual in that, although the hum level is high, the subscriber's television receiver is not affected.

More commonly, hum is caused by a corroded or loose seizure sleeve on a passive device, such as a directional subscriber tap, power inserter, directional coupler/splitter, amplifier housing, or cable splice.

Hum Measurement in the Time Domain

The spectrum analyzer is ideally suited for hum measurements because it can act as either a signal level meter with a built-in oscilloscope for observing the amplitude modulation, or it can be act as a sort of waveform analyzer for direct observation of the carrier's video modulation. The signal-level meter mode is used for hum measurements on unmodulated carriers, and the waveform analyzer is used for the measurement of hum on modulated carriers.

Naturally, it is desirable to read the interference with the modulation on to keep subscribers happy. Most of the time this is possible. But program modulation sometimes interferes with the hum measurement by causing transients on the analyzer's display, making hum look worse than it is. So if the readings are too close to the 3% limit with the modulated carrier measurement, it is necessary to measure the interference on an unmodulated carrier.

The Spectrum Analyzer as a Signal Level Meter with an Oscilloscope

Hum measurements demonstrate the versatile nature of the spectrum analyzer. Up to this point, the analyzer has been used to measure signals in the frequency domain. But hum is AM that must be observed in the time domain. Here is how the spectrum analyzer can be adjusted to make time-domain demodulation measurements.

TIP	**The spectrum analyzer can display the carrier's demodulated waveform in the time domain, just as an oscilloscope can.**

Take a look at Figure 116. This simplified block diagram for the spectrum analyzer shows its two basic operating modes. In (a), as a spectrum analyzer with frequency display, the input signal is mixed over a frequency range dictated by the analyzer's span setting which is set by the sweep voltage applied to the local oscillator. The signal at each frequency point is filtered, detected, and displayed in its rightful place along the display by the same sweep ramp applied to horizontal CRT input. The display is amplitude versus frequency.

A signal-level meter examines one signal at a time. It does not sweep in frequency unless you turn the tuning knob. In the spectrum analyzer, the sweep generator is turning its tuning knob all the time! If the sweep is disconnected from the local oscillator, as shown in (b), then the analyzer is tuned to only one frequency. This is called fixed-tuned mode, or, because the horizontal display is for only one tuned frequency, a better name is zero-frequency span mode.

TIP	**When the analyzer is fixed tuned, or, in zero span, the display becomes a time domain picture of the video wave form, like an oscilloscope.**

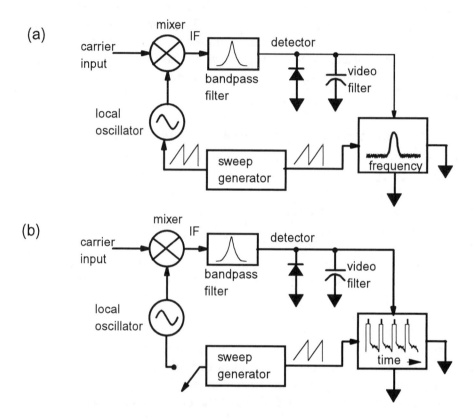

Figure 116. Spectrum analyzer block diagram to make (a) frequency domain measurements, and (b) amplitude modulation measurements in the time domain.

The analyzer's detector can now be used to demodulate the single carrier the analyzer is tuned to. If the IF or resolution bandwidth and video bandwidths are set wide enough, the diode detector tracks the rise and fall voltages of the amplitude modulation of the visual carrier. The sweep generator now provides a horizontal drive for the display that is scaled to time, not frequency.

In zero span the analyzer becomes a signal level meter with a built-in oscilloscope for displaying the demodulated waveform of the carrier, including the hum AM. To illustrate, observe the analyzer's CW calibration signal in the frequency domain and in the zero span modes.

Example 44. Use the spectrum analyzer in zero span to observe the calibration signal.

Connect the calibrator signal to the analyzer input. Set the analyzer to the calibrator's center frequency in a span of 1 MHz, and let the analyzer's resolution andand video bandwidths automatically adjust themselves.   You should have a display as in Figure 117.

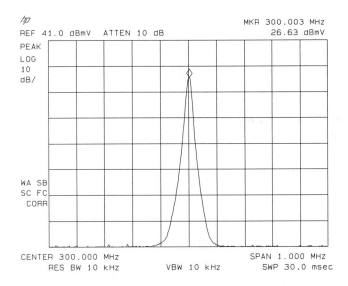

Figure 117. Analyzer's calibrator signal in the frequency domain.

Center the signal and set the analyzer to zero span by selecting a frequency span of 0 Hz. The analyzer is not sweeping over frequency any longer, but is tuned to the frequency that was at the center of the display before you switched to a span of 0 Hz. The straight line across the display represents the amplitude of the carrier. It does not change amplitude because the calibration signal is not modulated. See Figure 118.

Now the horizontal display is scaled for time. If you activate a marker, its readout is time instead of frequency. Two markers read the time between them. The analyzer's sweep function now controls the horizontal sweep just as an oscilloscope.

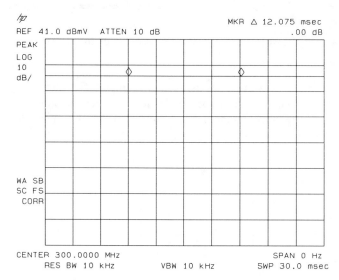

Figure 118. Calibration signal in the demodulated in the time domain.

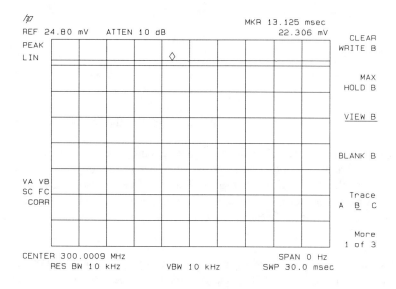

Figure 119. Vertical scale is set to linear, or voltage.

Finally, set the analyzer to read in voltage amplitude. This is important because to measure hum, the percent modulation must be a comparison of voltage levels, not dB. Select the linear mode from the amplitude controls of the analyzer. The signal level drops from where it was in the dB scale because the analyzer display has translated the reference level in dBmV to voltage. Move the signal up near the top using the amplitude control. The marker amplitude is in volts. See Figure 119.

For the hum level to be measured accurately, the voltage scale must have 0 volts at the bottom of the display; otherwise the amplitude ratios will be incorrect. To assure yourself that the base line on the analyzer display is 0 volts, pull off the input and see where the trace line goes. This is illustrated in Figure 119 in another trace along with the signal, although the base line may be difficult to see because it lies on the bottom graticule.

TIP   **Use a linear scale and voltage units to measure hum.**

CAUTION   **Make sure the bottom graticule is 0 volts.**

Measuring Hum on a Modulated Carrier

The measurement objective is to get an analyzer display to look like the picture in Figure 114, and then take amplitude readings and calculate hum. The next example shows how.

Example 45. Measure hum on a modulated carrier.

Center the carrier in the frequency display as you did the calibration signal in the last example. Set the resolution bandwidth to 1 MHz, the video bandwidth to 1 MHz, and the sweep time to 20 ms. Now adjust the amplitude for voltage by setting the amplitude to linear. Bring the signal to the top of the display. You may use the single

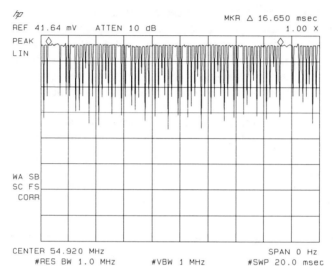

Figure 120. Demodulating the visual carrier.

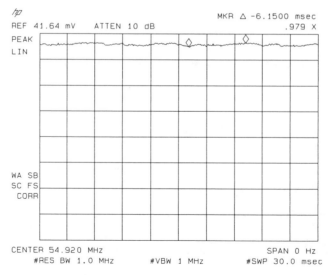

Figure 121. Increase of the sweep time eliminates the luminance "noise."

sweep mode to freeze the display. Figure 120 shows two vertical intervals at the markers. The marker readout is the time between them, 16.650 ms. The inverse of this is the vertical interval frequency, about 60 Hz.

This display is not adequate for the hum measurement. There is no clear way to determine the peak-to-peak variation of the sync pulse peaks because the analyzer is displaying some of the luminance modulation. This is because the analyzer is sweeping so fast that a sync pulse is not detected at each horizontal point. The analyzer detector is a peak detector, so that even one sync pulse is displayed if encountered. To remedy this, set the sweep time to 30 ms. Figure 121 shows the result. If this does not work for your analyzer, lengthen the time until all the points are on the same line.

Now the hum can be measured. Place a marker on the highest point of the display and a second marker on the lowest point. The first marker is the "peak" value, as shown in the hum definition diagram of Figure 114. The difference between the marker amplitudes is read out as a linear value less than 1.0. The peak-to-peak variation is the readout subtracted from 1.0. In this case the "peak-to-peak" value in Figure 121, is 1.0 − 0.979, or 0.021. Multiply this by 100 to get the percent hum interference, 2.1 %.

How much of this hum is due to the carrier's modulation and how much is power-line related? It is not possible to answer this question from the procedure. The only way to make sure is to measure the hum on the unmodulated carrier. But one indicator is to repeat the measurement of Example 45 several times and average the values. For example, in repeating the test, the values for the marker differences are 0.973, 0.981, 0.980, 0.975, and 0.979. Subtract each from 1.0, add these values, and divide by the number of samples, five, to get the average, that is, $(0.027 + 0.019 + 0.020 + 0.025 + 0.021)/5 = 0.0224$ or 2.24%. Wider variations over time would indicate that modulation is interfering with your measurement. In this case, the 2% hum level gives confidence that the modulation did not interfere. If it had, turning the modulation off would be the only way to confirm the hum level.

TIP

A consistent 2% level, or less, hum measured on a modulated carrier is probably accurate.

Sync-Suppressed Scrambled Channels

One special case of modulation-on hum testing is the sync-suppressed scrambled channel. Since the sync pulse is the amplitude value required to determine hum levels, you can measure the carrier after the sync has been restored by the de-scrambler at the subscriber terminal. Another way is to use the analyzer as in Example 45, but with extra long sweep times to let each display point pick up a vertical interval level.

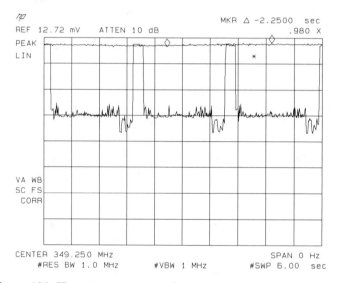

Figure 122. Hum measurement of a sync-suppressed scrambled channel.

Figure 122 shows the hum measurement. The sweep time was increased until every display point caught a vertical interval, about 6 seconds full sweep. The time it takes your analyzer to fill in the trace depends upon its total number of trace points. Experiment. The figure also shows the scrambled carrier vertical interval syncs when the analyzer is swept at a 5 ms/division rate. The marker readout shows a hum level of 2%, consistent with the earlier example from the same system.

Measuring Hum on the Unmodulated Carrier

With the modulation off, no sync pulses need to be demodulated, and the analyzer's video bandwidth, the equivalent of a low pass filter in the signal level meter, is narrowed to 1 kHz to smooth the short term amplitude variations. This makes it easier to pick out the amplitude variations. The sweep time is set to the NCTA-recommended 5 ms/division. Other than these settings, the measurement is identical to that of Example 45. Figure 123 shows the measurement made on an unmodulated visual carrier.

The reading of 0.993 translates to a hum level of 0.7%. Several readings of this carrier vary only one or two tenths of a percent, giving confidence to this reading.

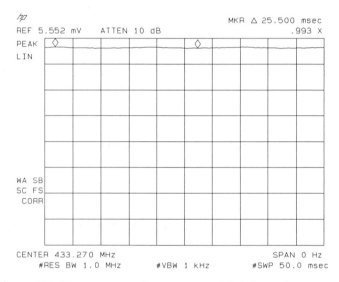

Figure 123. Hum measured on an unmodulated carrier.

FFT Snapshot May Help Diagnose the Source of Hum

Knowing how much hum comes from what frequency can give you a clue as to its origin, as discussed earlier. In fact, the frequency may not be power-line related, in which case you would be looking for nonpower supply causes. The FFT can help in this diagnosis. Figure 124 shows an FFT snapshot of an unmodulated carrier. The display indicates only significant 120 Hz content, which could point to a bad filter capacitor or low line voltage rather than a corroded connector or a faulty power supply diode.

The spectrum analyzer FFT function is not usually specified tightly enough in amplitude to allow it to be used for the measurement of hum levels.

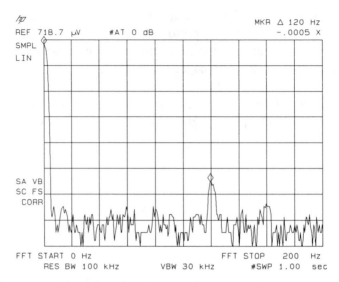

Figure 124. Dissection of the harmonic content of the hum interference with the FFT function.

  Use FFT for diagnosing hum source but not for the hum percent reading.

Hum Measurement Accuracy

The hum measurement is absolutely a relative measurement. Unlike the relative measurements used up to this point, the analyzer's amplitude scale is set to the linear mode to view the voltage responses of the carrier modulation, rather than the logarithmic mode. The specification of accuracy for the linear mode in spectrum analyzer data sheets is usually expressed in terms of percent uncertainty relative to the reference level.

Measurement Conditions Critical

More important than calculating the uncertainty of any specific measurement are the conditions under which the measurement is made. If the modulation is on, the contribution of uncertainty by the spectrum analyzer to the measurement is not critical because the data being measured is subject to the program modulation.

TIP

If hum is measured with the modulation on, hum levels greater than 3% need testing on an unmodulated carrier to be reliable.

For example, if the measurement gives greater than 3% hum, there is significant hum on your system to warrant retesting on an unmodulated carrier to determine a more accurate number. If you measure less then 2% hum with the modulation on, you do not need to retest, just apply the accuracy discussed below. Trust your own judgment, and your understanding of the characteristics of your plant for values between 2% and 3%.

Linear Uncertainty

When testing hum on unmodulated carriers, the major source of uncertainty is the analyzer's display fidelity, just as in all other relative amplitude measurements. Hum uses the linear amplitude scale. The typical spectrum analyzer data sheet in Appendix C shows the linear accuracy specification under the heading: Amplitude Specification, Display Scale Fidelity. It is specificed as ±3% of reference scale.

This value is the same as the specification, and as a general rule the analyzer has to be about ten times better in performance to make a reliable and repeatable measurement. Fortunately, two attributes of the spectrum analyzer help ease the guilt when calculating the accuracy in the face of such a high uncertainty. First, are the conservative specifications most manufacturer's place on their analyzers. The 3% value represents the worst case, under the worst conditions, of all the analyzers tested. Your analyzer is probably much more accurate. Second, nonlinear response tends not to change rapidly with a change of input level. This means that the uncertainty is proportional to the amount of peak-to-peak voltage variation. Another way to look at that is that the uncertainty is a function of the measured hum level. The higher the hum, the higher the measurement uncertainty. Here are typical specifications for hum accuracy: ±0.4% for ≤±3.0% hum, ±0.7% for ≤±5%, and ±1.3% for ≤±10%.

Example of Hum Accuracy

In Example 45 the hum level was measured as 2.1%. From the guidelines above, this value uncertainty is ±0.4%, or a hum from 1.7% to 2.5%. In other words, when measuring hum with the carrier unmodulated, the uncertainty is about 15% of the hum level.

Summary of Procedure

Here are the important points to remember about using a spectrum analyzer to measure hum:

- Make the measurement on the visual carrier or pilot tone.
- Hum on one carrier usually represents hum on the entire system at that point in your distribution branch, and over all the carriers.
- Measurements can be done with or without carrier modulation.

When Testing with Carrier Modulation on

- Hum levels above 3% must be retested on an unmodulated carrier.
- Use a resolution and video bandwidth of 1 MHz to capture all the sync pulse peaks.
- Set sweep time to guarantee one peak per point (at least 30 ms for 400 points), but experiment.
- Don't bother computing the uncertainty of the hum reading because modulation obscures the reading.

When Testing with Modulation off

- Set the analyzer's video bandwidth to 1 kHz to act as a low pass filter to capture only the low-frequency disturbances.
- Set the sweep time to 50 ms.
- Multiply the hum reading by 1.15 to account for the analyzer's uncertainty.

Selected Bibliography

- Goehler, Randy. *Personal communication*, March 1994.
- *NCTA Recommended Practices for Measurements on Cable Television Systems*. 2nd ed. rev., October 1993.
- Webb, Jack. *Understanding and Using "On-channel" Tests to Measure C/N and Hum*. Communications Technology Publications Corp., June 1993.

11

Leakage, Co-Channel, and Ingress

Overview

Signals found where they should not be are costly and dangerous. Costly because interference within your system causes complaints and unhappy subscribers. Dangerous to life and limb because the signals that leak from your system that are greater than 15 μV/meter can cause interruptions to communications for air navigation and jeopardize public safety. Of the interference that intrudes on your system, co-channel is the most common, and most easily traced back to its source. Less well-defined interference with program material, called ingress, may be from over-the-air broadcasters and paging systems, and may be tougher to find. Leakage and ingress sometimes go together. Interference may be injected into a system by the same breach in the system's cable drop that leaks. This chapter shows you how the spectrum analyzer can help you find signals trespassing into or out of your system.

Leakage – Signals Getting Outside the System

Signal leakage is caused by your cable's signals radiating into the air from an abnormal break in the distribution hardware. Partly because the term "radiation" sounds threatening to subscribers, it has been replaced by the term leakage.

FCC Regulations

The fear of harm coming to aeronautical navigation and communications is central to the regulation and enforcement of leakage. FCC studies concluded that leakage has to be measured on a power summation basis to make testing procedures practical, yet decisive. The power summation is a procedure/specification known as the cumulative leakage index, or CLI. This index is a figure of merit for a system based upon the power summation of all the leaks over a specified level.

Leakage Measurement Without the Fly-Over

From the air over a system, where interference with aeronautical communications is most likely, the cable system acts like a large antenna array. But flying sophisticated antenna, receiver, and data recording equipment over a system is an expensive procedure and may not be repeated frequently enough to measure the system under all conditions. The FCC CLI procedure allows data to be gathered from ground measurements during normal maintenance. If sufficient leakage information is gathered, CLI can be calculated without the fly-over.

TIP

CLI computation can be made from ground-based measurements provided enough data is taken, using approved procedures.

How CLI Represents Power Summation

One method of calculating CLI is as the sum of all the leaks of a system measured at a height of 9,800 feet (3,000 meters). Figure 125 shows this graphically. If the leakage at measurement point, L, is over a specified level, it is recorded along with its position, given as distance, r, from the CLI calculation point. The leakage at point L is measured on the ground in µV/meter from 10 feet away. The field strength in µV/meter is read as microvolts per meter.

The Measurement in Brief

To Determine the Leakage Level at a Specific Cable Location

1. To determine the leakage level at specific cable location. Select a calibrated dipole antenna that has current calibration data.
2. Place the dipole antenna 10 feet from the suspected leakage source, or measure the distance in feet.
3. Set the analyzer for 300 kHz resolution bandwidth and video bandwidth and 6 MHz span.
4. Center a signal that you know is unique to your cable.
5. Let the trace to maximum hold and gather the highest response.
6. Use the marker to read the peak leakage signal.
7. If the antenna was more than 10 feet from the leak, correct the reading for distance.
8. Compare the value with the leakage compliance specifications.
9. Record these findings along with your location to help with CLI computation.

Ingress and Co-Channel

1. Zoom in on an ingress or co-channel signal with a span of 50 kHz, a resolution bandwidth of 1 kHz and a video bandwidth of 10 Hz.
2. Use two markers to look at the level differences between the nearest visual carrier and the ingress or co-channel in dB.
3. For ingress, use the analyzer's demodulator and frequency counter to help identify the source.
4. If evidence is required, record your results using plots, prints, trace storage, audio recording, or videotape recording.

TIP **Levels greater than 50 μV/m should be recorded for the CLI calculation, but it is a good idea to keep track of the leaks greater than 20 μV/m for maintenance purposes.**

This value is squared to represent its power contribution, and then divided by the square of the distance to the measurement point, R, to give it a weighting for position. A leak that is further away, that is, where R is greater, has less influence on the cumulative leakage. The

CLI number has no units; it is a figure of merit used to measure a system against the FCC specification.

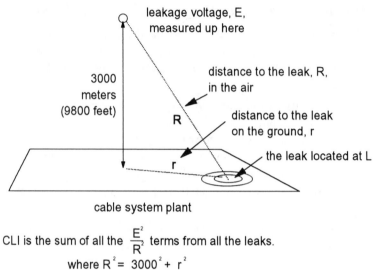

CLI is the sum of all the $\dfrac{E^2}{R^2}$ terms from all the leaks.

where $R^2 = 3000^2 + r^2$

Figure 125. Concept of cumulative leakage index.

> **TIP**
>
> **CLI is based on the data from all the leaks greater than those specified by the FCC. But every leak plugged reduces CLI.**

If this confuses you, don't worry. The important lesson is the need for gathering ground-based leakage measurement data in your daily routine and how it is related to CLI. Recording leakage data on the ground will keep you off the 9,800-foot high cherry pickers, or out of the fly-over plane.

Ground Measurement Specifications

FCC Rules and Regulations, Part 76.605 (a)(11) offers specific field strengths over frequencies and distances for leakage using a simple tuned dipole antenna. Following these guidelines is the first step in ensuring that your system meets the CLI limits. Correcting the leak is the second step.

Table 7. Radiation limits for leakage when measured on the ground.

Frequency MHz	Field Strength μV/meter	Antenna distance from leakage feet
<54	15	100
54 to 216	20	10
>216	15	100

Since inspection usually concentrates in the 108 to 137 aeronautical band, you will probably want to test at least one carrier within one of these bands.

TIP   **At the very least, test for leaks between 108 and 137 MHz.**

Measuring Leakage with the Spectrum Analyzer and Dipole Antenna

When you are in the field making measurements with the spectrum analyzer, it is a good idea to have a dipole antenna handy for making leakage measurements. This is especially true when visiting older plant installations, where more leaks are liable to be found. The analyzer's ability to see over broad spans helps find and measure leaks quickly. After all, your time is money.

Dipole Antenna

The dipole antenna, used for its simplicity and comparative ease of calibration, looks like the capital letter T, where the top crossbar consists of two telescoping whips, or rod antennas. The voltage at the antenna's output connector represents a specific level of radiation which is measured as voltage over so many meters of air. The units of radiation, as in Table 7, are read as microvolts per meter (μV/meter). A microvolt is one millionth of a volt, or 1×10^{-6} volts. A small number, but the compliance specifications are aimed at preventing radiation from leakage going higher than 10 μV/meter at the radio or transceiver subject to interference.

You can build your own antenna cheaply, but there are a number of well-designed and thoroughly documented antennas on the market that can make your job easier from the start. Here are some important antenna features to consider:

- Frequency range that covers your system's current and future bands.
- Documentation that includes calibration information, that is, traceable to the National Institute of Standards and Technology (NIST).
- Manufacturer who calibrates the antenna at a reasonable price in a short time.
- Table of antenna output levels for the FCC specifications, by frequency.
- Built-in amplifier for improving sensitivity when measuring at longer distances.

TIP

It is good practice to send your antenna to its manufacturer for calibration periodically, especially if the antenna has been treated roughly.

Units of Measure

The antenna output is a voltage on a 75Ω cable. The voltage represents the radiated power being picked up by the antenna. A conversion factor, which is different for each frequency, is provided by the antenna manufacturer. This value is required to convert the spectrum analyzer signal level to field strength. Manufacturers may or may not provide you with a conversion factor that includes the limit specified by the FCC; you are responsible for proper calibration of your test equipment. Here is an example.

Example 46. Dipole correction factors.

Your shop has a dipole antenna with the antenna correction factors displayed in Table 8.

The right-hand column gives the FCC limits as read on a spectrum analyzer or signal level meter in dBmV. The antenna manufacturer has combined the FCC limits and his antenna corrections all into one value for each frequency. If the analyzer reads over −25 dBmV on a leakage signal at 108 MHz from the antenna placed 10 feet from the cable, then the leak is over the allowed 20 μV/meter limit at that frequency.

Table 8. Antenna correction factors

Channel	Frequency MHz	Distance from cable feet	Maximum SLM dBmV
FM	108.00	10	−25.0
A	121.25	10	−26.0
C	133.25	10	−26.8
11	199.25	10	−30.3

To determine the levels equivalent to the 50 μV/m limit for CLI data collection, just add the dBmV equivalent of a change from 20 to 50 μV/m in dB to the numbers in the right-hand column. The formula is found in Table 10 of Appendix A, the conversion of volts to dBmV. Convert each μV level to dBmV and subtract as follows: 20 μV = 20 × 10^{-6} V, and dBmV = 20 log (20 × 10^{-6}/10^{-3}) = −33.98 dBmV. Likewise, 50 μV = −26.02 dBmV. The increase is −33.98 − (−26.02) = 7.96 dB. The 50 μV/m values become −32.96, −33.96, −34.76, and −38.26 dBmV.

CAUTION

Not all antennas provide a table of leakage limits in units read by the spectrum analyzer. If not provided, you will have to calculate them.

If the antenna documentation includes only unitless antenna correction factors, or dBmV per μV/meter, it is your responsibility to calculate the limits for the FCC specifications. References are available to help with these computations. But it may be far simpler and cost-effective to purchase a new, calibrated, and documented antenna. The majority of antennas sold today provide exact limits as in the example. It is important to have the antenna's calibration history when providing data for leakage performance.

Dipole Tuning

The antenna collects RF-radiated power along its length, a distance designed to be a half the wavelength of the frequency to be measured. The antenna correction factors assume that it is tuned. Practically, you may not see the readings change much when changing the antenna's length, but it is good practice to make the adjustment. If you test primarily one cable system frequency, only one length setting is required.

Your antenna should come with a table of lengths for each channel or frequency. If it doesn't, Figure 126 shows the general formula for dipole lengths and an abbreviated table. Note that the length, L, is the total antenna length, not the length of one of the rods.

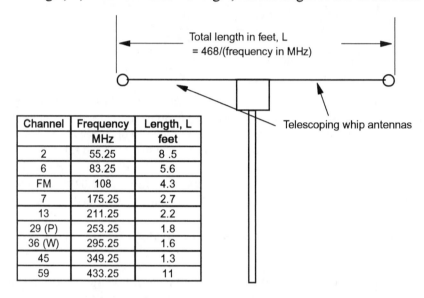

Total length in feet, L
= 468/(frequency in MHz)

Telescoping whip antennas

Channel	Frequency	Length, L
	MHz	feet
2	55.25	8 .5
6	83.25	5.6
FM	108	4.3
7	175.25	2.7
13	211.25	2.2
29 (P)	253.25	1.8
36 (W)	295.25	1.6
45	349.25	1.3
59	433.25	11

Figure 126. Dipole dimensions and length adjustments.

TIP

Some manufacturers may give the dipole dimension for each whip, not for the overall length.

Test Signal Selection

Select an appropriate signal to measure. A carrier or test tone can easily represent all the signals in a cable because few, if any, leakage mechanisms change the frequency response of the radiation bandwidth. Here are some other tips on selecting the signal to test.

- Avoid interfering broadcast or other over-the-air signals.
- Use a regular visual carrier or leakage test signal when possible.
- Avoid signals that could be generated by a neighboring cable system.

Prevent Analyzer Overload with the Input Attenuator

Measuring leakage and ingress usually means measuring low-level signals when a group of high-level signals are input into the analyzer's mixer. This means that the analyzer can be overloaded. Use the internal attenuator test as described in Chapters 4 and 8 to prevent gross measurement errors.

Comparison to Field Strength Meter

The field strength meter, also known as a signal level meter or signal leakage detector, is a simple receiver for measuring leakage. The spectrum analyzer, like a field strength meter is a tuned voltmeter, measuring small signals in the presence of large signals. One significant difference is in the way the signal levels are detected.

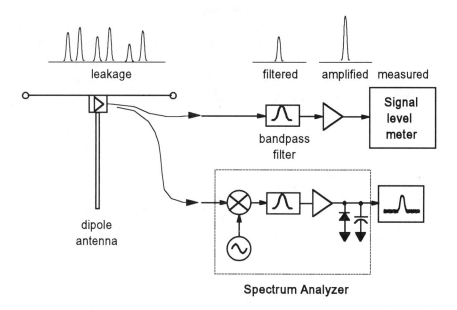

Figure 127. Spectrum analyzer and field strength meter block diagrams.

The field strength meter filters the input response and detects the RMS voltage of the leakage signal. The spectrum analyzer filters the input with its resolution and video filters, but the detection responds to the peaks of the leakage, not the RMS. The signals in the spectrum analyzer may be as much as 5 dB higher when compared to the readings of a field strength meter.

The same spectrum analyzer leakage reading can be 5 dB higher than it would be on a field strength meter.

Since the analyzer presents the worst-case leakage level, why not accept its reading as actual to be conservative? A bit more about this is said in the section on accuracy below.

Leakage Measurements

In field measurements, use the analyzer's broad frequency-measurement capability to look over the entire frequency range, keeping in mind the limitations placed on the measurement by the antenna's bandwidth. Figure 128 shows such a quick survey. The lower trace is the over-the-air survey, and the upper trace shows the cable spectrum. Both traces are made with the same reference level, but the radiation sweep was made with a narrower resolution bandwidth to improve the analyzer's ability to see small signals.

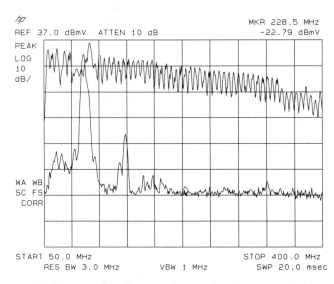

Figure 128. Survey of leakage and over-the-air transmissions.

From this display it is easy to see broadcast signals that could interfere with the leakage measurement. In this case, the only large signals are from commercial FM broadcasting. It is important for you to know your local over-the-air signals to prevent mistaking them for leakage, or mistaking leakage for broadcast signals.

How to Make a Measurement

Select a suspected leakage location on the cable and move as close to it as practical. If a drop site is nearby, it is the most probable source of leakage. Using a span of 6 MHz, and video and resolution bandwidths of 300 kHz, center the appropriate leakage signal. Set the correct dipole length for the frequency and place the antenna to within 10 feet of the cable if possible. Position the antenna parallel to the longest cable stretch and adjust the antenna for the maximum signal amplitude.

If there are a large number of high-level broadcast signals in your environment, test the analyzer for compression-type overload. Use maximum trace hold and a marker to measure the peak in dBmV or dBμV, depending upon the limit units your antenna provides. Compare the levels to determine if the leakage is larger than allowed.

CAUTION — If the antenna has a built-in amplifier, be sure its gain is included in the antenna limits. If the amplifier is not used, subtract the gain from the antenna's compliance limit value.

Whether the leakage level exceeds the limit set by your system's guidelines for preventive maintenance or not, record the readings, location, date, and time. Almost all leakage levels contribute to the computation of CLI.

Correct for Distance

If you cannot get the antenna within 10 feet of the cable because of access or geography, take a leakage reading anyway, then apply the following simple formula to convert the leakage reading into a 10-foot reading. The leakage drops as distance increases by 20 log (actual distance/10). For example, if the leak is 150 feet away and is measured at −45 dBmV/m, the leak at 10 feet would be −45 dBmV/m + 20 log(150/10), or −45 + 23.5 = −21.5 dBmV/meter. There are a number of optical gadgets on the market that allow you to determine the distance to the cable drop just by knowing the size of an object on the pole, such as the length of a trunk or feeder amplifier case.

TIP

When more than 10 feet from a leak, the reading is low by 20 log (actual distance/10) in dB. Take the reading and then add the distance correction.

Here are two examples that follow the recommended practices for measuring the leakage using a visual carrier 10 feet away and with a test tone 50 feet away.

Example 47. Measure the leakage from the cable of a television carrier signal.

Place the dipole antenna 10 feet away from the suspected cable hardware. Be sure there are no other cables, such as telephone or power cables, between the antenna and the television cable. Select a video carrier for a channel that is not broadcast in your zone, or likely to be carried by a neighboring cable television service. In this case the video carrier of channel 11 at 199.25 MHz is used. Tune the dipole's length for the frequency. The length of the total antenna (both whips) is 468/199.25 which is about 2.4 feet or 29 inches. Each rod antenna is set to 14.5 inches length. Set the analyzer to a center frequency of 199.25 MHz, and span to 6 MHz with 300 kHz resolution and video bandwidths. Set the trace to maximum hold, let it sweep a few times. Read the signal's amplitude using the marker. See Figure 129.

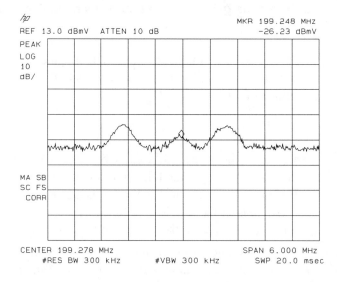

Figure 129. Using a video carrier for leakage test.

The leakage power is −26.23 dBmV. Look up the limit in the antenna's correction table found in Example 46. The limit is −30.3 dBmV which means that the signal is out of specification by about 4 dB (−26.4 −(−30.3)). A significant leak that needs to be repaired.

Example 48. Measure leakage from a distance greater than 10 feet.

A drop site is suspected of leaking. It is awkward to get close to the drop so you measure from the street, and you estimate the distance to the drop to be 50 feet. The cable test signal is a 108 MHz tone modulated carrier inserted at the head end for the purposes of detecting leakage. Connect the dipole, and use its internal amplifier. Set the dipole total antenna length to 468/108 = 4 feet 4 inches, or 2 feet 2 inches on each side.

Connect the analyzer, set the center frequency to 108 MHz, span to 6 MHz, and 300 kHz resolution and video bandwidths. Set the trace to maximum hold, let it sweep a few times, and read the signal's amplitude using the marker. See Figure 130. The level read is −21.09 dBmV, or −21 dBmV.

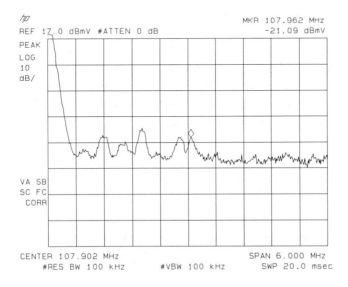

Figure 130. Leakage from a 108 MHz radiation tone.

Correct the measured level for distance. The correction for 50 feet is $20 \times \log(50/10)$ = 13.9 dB, so the leakage level for 10 feet is $-21 + 13.9 = -7.1$ dBmV. The antenna limit specification at 108 MHz, as in Example 46 is -25.0 dBmV. The signal is significantly higher, and must be reported and corrected.

Signals whose levels are even 10 dB below the limits as measured on the spectrum analyzer should be reported without delay. Even though a field strength meter may see 5 dB less signal in some cases, leakages of this level usually mean a break in hardware.

Accuracy

When you are in the field and waving the dipole around, you can easily see ±5 dB variations. The same is true for weather and temperature changes. So why quibble over a few tenths of a dB for accuracy? You probably need not for most leakage measurements, especially if you always take the highest readings, that is, err on the high side. Being conservative and using all the data in the power summation process of CLI keeps your system tight and safe.

Signal Ingress and Co-Channel

Where there is leakage, there can also be signals getting into the system. In fact, when you hear complaints about odd interference from your subscribers, such as citizen's band or mobile communications radio, suspect a break in cable hardware. The most common ingress, however, is co-channel. It rarely sneaks into the cable system through a leak, but enters through your head end off-air antenna site.

The Appearance of Co-Channel

Co-channel interference is one of the most obnoxious television types because it forms ghosts or heavy dark bars over the picture which may swim at random. It can appear as a second channel dimly lit against the strong channel, usually out of sync so the boarders are skewed diagonally. If the co-channel is very close in frequency and sync, it may look a lot like CTB.

Figure 131 is a simulation of one view of co-channel distortion.

Figure 131. Simulation of co-channel effect on the television picture.

Co-Channel Sources

Over-the-air broadcast transmitters are licensed in zones to protect each other from interference under most conditions. In your cable system, whose job it is to import distant over-the-air channels, tall antenna towers and high-gain receiver amplifiers are used to gather these signals. They also gather other channels. Since an antenna is not perfectly directional, when aimed at a transmitter 50 miles away, it can also receive another transmitter from a different direction. Since the channels are within 10 kHz of one another, the undesired channel cannot be filtered out. However, a second antenna with a phase correction device is sometimes used to reject the co-channel right at the reception point itself. The spectrum analyzer is useful in setting installation and maintenance of these systems because its wide distortion-free amplitude range allows you to see the suppression of the co-channel as a result of the phase combiner.

Other Ingress

Other ingress have no hard and fast rules about how they appear to your subscriber, nor how they enter the system. When you get a complaint, the spectrum analyzer is your best tool for discovery and diagnosis, however.

FCC Regulations

There are no specific guidelines for ingress and co-channel, so it is common practice to use the rules established for coherent distortion, CTB/CSO, which say the levels should be 53 dB below the nearby visual carriers. Even though the measurement of co-channel requires a wide amplitude range free of analyzer distortion, the dangers of overload are less than they are for CTB/CSO because co-channel and ingress are rarely related in frequency to the system channel carriers. This means that the analyzer does not distort the ingress amplitude when overloaded except for compression. Follow the rules for overload protection in Chapter 4.

Nature of Co-Channel

Co-channel, by definition is found within ±10 kHz from its associated visual carrier. This means that the 300 kHz resolution bandwidth so often used in cable measurements cannot be used; the co-channel signal cannot be resolved. A 1 kHz bandwidth in a 50 kHz span gives you the required resolution, while improving the ability of the spectrum analyzer to see low-level signals even with the video carrier on screen. The video bandwidth is set to 10 Hz to filter out the carrier's modulation. Then small interference signals can be resolved between the video carrier's 15.75 kHz sidebands.

TIP	**Resolve co-channel with a resolution bandwidth of 1 kHz in a 50 kHz span. Set the video bandwidth to 10 Hz.**

Co-channel and other ingress may change with weather effects such as inversion layers over water, and other changes due to the time of day or time of year. The analyzer can help see these patterns over time when co-channel snapshots are taken throughout the year.

Example 49. Measure co-channel.

Complaints of interference described as diagonal bars moving across the screen have been coming in from subscribers about channel 6 on your cable. Measure the co co-channel.

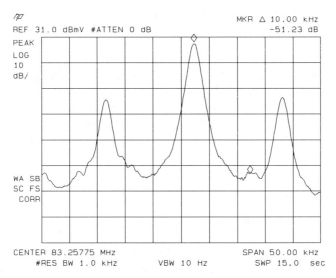

Figure 132. Co-channel measured on channel 6.

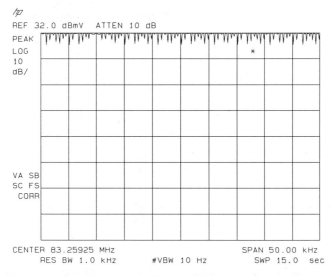

Figure 133. Measure the carrier level after co-channel has been found.

There is no need to go out to the subscriber tap. Co-channel is usually uniform throughout the system since its source is in front of the head end.

Select a center frequency of 83 MHz and set the span to 50 kHz, the resolution bandwidth to 1 kHz, and the video bandwidth to 10 Hz. Let the analyzer sweep one full sweep and use the markers to measure the co-channel as in Figure 132. The co-channel is 51 dB below the visual carrier. At this low level it is unlikely that co-channel can be seen by the subscriber. But if complaints continue, monitoring channel 6 over a 24-hour period might reveal cycles with temperature or weather, causing periodic increases.

The visual carrier level is measured with such narrow resolution and video bandwidths that its level is not accurate. This is suitable for quick looks as in the example, but for full documentation measure the co-channel level relative to the visual carrier level as measured for coherent distortion. To measure the full carrier level, open the resolution and video bandwidths to 300 kHz while in the same span where the co-channel is measured. Figure 133 shows how this looks. The carrier peak is at the reference level, +32 dBmV. The visual carrier is about 3 dB higher than it registered in the example measurement. This means that the co-channel is down about 54 dB from the visual carrier peak.

Finding and Identifying Other Ingress

The analyzer has other features that can help detect and identify ingress:

- AM/FM demodulation and speaker or headphone output.
- TV monitor output for recording the spectral response of an intruder.
- Phone jack for recording the audio demodulated from an interfering signal.
- Plot and print capability of the display to record all the spectrum data and a time/date stamp.
- Frequency accuracy markers to determine if the interference is a harmonic or subharmonic of a known local over-the-air signal.

> **TIP**   **Ingress can be identified using the analyzer's demodulator to listen to the ingress audio.**

Using the Demodulator

Listening to the interfering carrier is the most common method of identification. The demodulation circuit of the analyzer taps into the analyzer's IF circuitry while the analyzer is either in zero span, or while the sweep is stopped at a marker for a period called the dwell time. An AM or FM demodulator receives the IF signal and the audio is fed to an onboard loud speaker. Figure 134 shows the display during the demodulation of an FM signal. The dwell time is the time the analyzer's sweep is stopped for listening. (Sorry, you can't hear anything!)

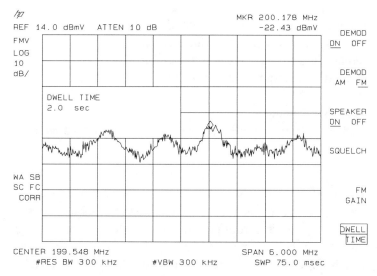

Figure 134. Using a demodulation feature to listen to an FM channel.

Accuracy When Measuring Ingress

The accuracy of ingress depends upon the intended use of the data. Minimizing co-channel depends more on the repeatability of the analyzer's measurement than on its accuracy. Taking analyzer snapshots of co-channel performance, such as in Figure 132, on channels you know are susceptible is the best preventive maintenance practice. Catch trends before they become complaints.

With other ingress, the analyzer's frequency accuracy is usually sufficient to trace the local transmission source signal.

Summary

Leakage

The analyzer cannot make CLI measurements, but, like any signal level meter or field strength meter, it contributes to the pool of data required for the ground measurements. One key to keeping leakage measurements simple is to have a dipole antenna whose correction factors are provided by its manufacturer. Antennas specified with the FCC-equivalent limits over frequency are valuable in saving you time and effort. Other considerations when measuring leakage:

- Place the antenna a known distance from the leak. If the distance is greater than 10 feet, then make a correction to the leakage level.
- Always adjust the antenna length for the leakage frequency.
- Measure with the spectrum analyzer in 300 kHz resolution bandwidth and 6 MHz span.
- Select a test signal unique to your system.
- Don't worry much about detailed accuracy analysis; worst case measurements are safest.

Ingress and Co-Channel

The most important ingress is co-channel, which is usually generated at your antenna tower, so it can be monitored as part of periodic preventative maintenance procedures. Measure co-channel with the analyzer in 50 kHz span, 1 kHz resolution bandwidth, and 10 Hz video bandwidth. Co-channel must be 50 dB or so below the visual carrier.

Other ingress can be identified by using analyzer features such as an AM/FM demodulator and frequency accuracy. Record ingress characteristics, including demodulated audio, to help settle interference disputes.

Selected Bibliography

- *Cable Television System Measurement Handbook*. Hewlett-Packard Company, Literature No.5952-9228, Santa Rosa, CA, January 1977.
- Chesley, Theodore R. *Signal leakage for Installers*. Communications Technology Publications Corp., January 1994.
- Dickenson, V. C. "Basics of Cumulative Leakage Index", Signal Leakage Supplement, *Communications Engineering & Design*, June 1989.

- Gordon, Robert, and Beu, Frederick E. "Monitoring & Measuring Signal Leakage", *Student Manual*. CATV Technical Services Division, Orion Business Services, Inc., Washington, D.C., June 1990.
- Hartson, Ted. "The History of Signal Leakage: An Unauthorized Autobiography", Signal Leakage Supplement, *Communications Engineering & Design*, June 1989.
- James, Brian. "Aeronautical Band Access: The New Regulations", Signal Leakage Supplement, *Communications Engineering & Design*, June 1989.
- *Final Report to the Federal Communications Commission,* Advisory Committee on Signal Leakage, Exton PA: Society of Cable Television Engineers, November 1981.
- *NCTA Recommended Practices For Measurements on Cable Television Systems.* 2nd ed. rev., October 1993.
- Shimp, Dick. *Signal Leakage Calibration, etc.*, Communications Technology Publications Corp., January 1993.
- Windle, Steve, and Vendely, John. *A Review on Leakage Detection and Measurement using Wavetek Gear- Part 1.* Communications Technology Publications Corp., July 1991.
- Wong, John. "Scrutinizing CATV Operators: The FCC Gets Tough", Signal Leakage Supplement, *Communications Engineering & Design*, June 1989.

Audio and Video Measurements

12

Overview

For the standard spectrum analyzer, audio and video measurements are conveniently made but are not as precise as those made by specialized equipment such as waveform analyzers, audio analyzers, broadcast quality modulators, and spectrum analyzers adapted for cable television and video measurements. This chapter shows you how to get a quick look at your cable system's audio and video without a lot of fuss. The quality of these measurements are sufficient to alert you to problems or to confirm customer complaints.

In this chapter you'll learn about:

- FM Deviation
- Setting digital signal levels
- Depth of modulation

FM Deviation

Each channel's audio signal is frequency modulated (FM) and summed with the visual carrier before transmission. Any time the baseband video signal is processed, such as for some satellite and AML transmissions, monitoring the FM quality prior to distribution at the head end is a good preventive maintenance practice. Checking your modulator's deviation warning light or meter is preventive maintenance only if the modulator has been calibrated off-line. Turning the deviation up until the light goes on, and then backing off a quarter turn, does not constitute calibration! Calibration can be done by the manufacturer or by you with a broadcast-quality deviation monitor or a signal generator and spectrum analyzer. This section:

+ Reviews FM concepts.
+ FM deviation calibration of modulators.
+ Provides a method for checking a channel's deviation using program audio.

TIP

A head end modulator's FM deviation warning light or meter is only as good as the calibration performed on the modulator.

FCC Rules for FM Deviation

There are no strict compliance rules for FM deviation in your cable system. FM is acceptable if the audio of each channel is clear and uniformly loud from channel to channel; when the subscriber is surfing through the channels, no channel's audio booms out or gets lost.

In each channel the vestigial filter eliminates the much of the lower AM visual sideband, minimizing the bandwidth and leaving room for the lower adjacent channel's audio carrier. The FCC mandates that the aural carrier appear 4.5 MHz above the video and that its peak deviation be less than ±73 kHz when stereo and second audio programming, or SAP, is transmitted. Monaural peak deviation is set to ±25 kHz. The amplitude of the aural carrier in a cable system is several dB lower than specified for broadcast transmission. The reason–fewer intermodulation products and less chance of adjacent channel interference in a crowded cable system.

The Measurement in Brief

FM Deviation from Program Audio

1. Use the analyzer's calibration signal.
2. Set the span to 400 kHz and the resolution bandwidth to 100 kHz.
3. Offset the center frequency to position a linear portion of the filter sideband at the center.
4. Mark the sideband with two markers set 50 kHz apart and record their amplitude difference in dB.
5. Connect the cable signal.
6. Center the audio carrier, and then offset it from the center by the same amount as in Step 3.
7. Place the analyzer in zero span.
8. Read the FM deviation as amplitude modulation, in dB, scaled with the amplitude recorded in Step 4. This is peak-to-peak deviation.
9. Peak deviation is half the value calculated in Step 8.

Depth of Modulation from Program Video

1. Center the visual carrier with a 300 kHz resolution bandwidth and a 300 kHz video bandwidth.
2. Select zero span.
3. Bring the signal near the top of the display and change the vertical scale to linear (voltage).
4. Set the sweep time for 5 µs, or the minimum your analyzer allows.
5. Bring the top of the signal to the reference level.
6. Collect the minimum trace responses in a trace memory. If your analyzer does not have a trace minimum function, use a display line to mark the minimum observed swings.
7. The depth of modulation is the minimum deflection as a percent of the total display amplitude. Calculate the results. The value should be 87.5% or less.

TIP

Few complaints about sound quality is your reward for good FM signal quality maintenance.

Symptoms of Poor FM Deviation

If the modulator produces too high a deviation, the subscriber suffers excessive volume on the overdriven channel. If it is bad enough, the sound distorts. Too little deviation, and the volume is low. Compensation at the television for low volume makes the sound loud but muffled and noisy.

Your ear can hear the worst extremes of these symptoms, but the spectrum analyzer can do a better job of quantifying them. But first a review of FM.

FM Concept

Television uses an FM sideband carrier because:

- ♦ It adds little power to the channel band.
- ♦ It does not interfere with the amplitude modulation of the video carrier.
- ♦ Its power does not increase and decrease with signal modulation.

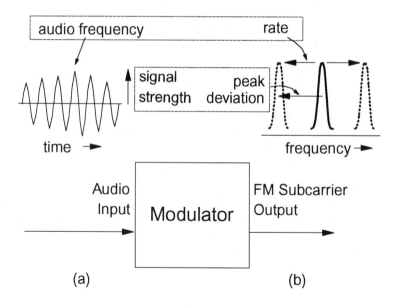

(a)   (b)

Figure 135. How the audio signal is transformed into a carrier by the modulator.

In FM the audio signal strength moves the modulator's output in frequency away from the carrier. The stronger the audio, the further the FM carrier swings away from its assigned frequency. The rate at which this swing occurs is the frequency of the audio signal. A 500 Hz audio tone makes the aural carrier move back and forth at a 500 Hz rate. Figure 135 illustrates this concept.

In the figure, an audio signal at (a) has frequency and strength. The modulator converts strength to the FM signal's deviation (b), or how far the signal swings in frequency from its carrier. The FM carrier does not change amplitude in this process. This quality of the FM scheme makes it a good companion subcarrier for the visual carrier, whose primary modulation is AM; the FM signal contributes negligible amplitude modulation to the channel band.

An FM signal does not change power level as it is modulated, so it does not interfere with the visual carrier's amplitude modulation.

FM bandwidth is specified by its maximum deviation, called peak deviation. Too high a deviation causes the television to be overdriven, too low causes low volume. How is deviation measured?

Bessel Null Concept

Program audio is too random to set a modulator's deviation. Using a controlled audio tone at a strength required by the modulator, the deviation can be set precisely. This is possible because FM behaves in a very predictable way when driven by a sine-wave audio signal. Curiously, the spectral response of an FM signal is precisely known for every unique ratio of peak deviation to audio frequency, sometimes called the modulation index. The FM sidebands are predicted by mathematical data called the Bessel functions. Among other things, Bessel function tables show that the FM carrier is zero, that is, has no amplitude, when the modulation index is 2.4048. There are books filled with Bessel tables, so there are many combinations of audio frequency and deviation where the modulator can be set precisely. Table 9 shows the most popular.

The table shows the modulation index for the null of the carrier and first set of sidebands. Figure 136 shows how the nulling looks on a spectrum analyzer monitoring the

Table 9. The first eight important Bessel values.

Times the carrier or sidebands go through a null	Ratio of peak deviation and modulation frequency (modulation index)		Modulation frequency in kHz for	
	Carrier null	First sideband pair null	25 kHz peak deviation	75 kHz peak deviation
First	2.4048	3.83	10.396	31.188
Second	5.5201	7.02	4.527	13.587
Third	8.6531	10.17	2.889	8.667
Fourth	11.7951	13.32	2.120	6.361
Fifth	14.9309	16.47	1.674	5.023
Sixth	18.0711	19.62	1.383	4.150
Seventh	21.2116	22.76	1.179	3.536
Eighth	24.3525	25.90	1.027	3.080

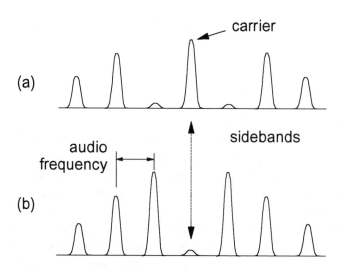

Figure 136. Spectrum analyzer showing (a) the carrier and sidebands, and (b) the Bessel null of the carrier.

modulator's RF or IF output. Note that the spacing between the sidebands is the audio signal's frequency.

TIP

Use the Bessel null technique to set modulator maximum deviation precisely.

To calibrate a modulator on the bench, connect the spectrum analyzer to the output of the modulator and tune to its output frequency, whether RF or IF.

Example 50. Set up a modulator for the first and eighth carrier null.

Set the analyzer's span for 30 kHz so that the audio sidebands can easily be resolved. Input an audio tone to the modulator at strength specified for 25 kHz peak deviation. Set the audio frequency to 10.396 kHz, and adjust the modulator for the appropriate 25 kHz peak deviation light or meter response. For 75 kHz peak deviation set the audio tone to 31.188 kHz.

Since the audio signal's input to the modulator are pre-emphasized, that is, boosted in strength at the higher modulation frequencies, you may wish to use an audio signal close to 1 kHz, where pre-emphasis is low. This way the modulator has no distorting effect on calibration. Table 9 shows that a 1.027 kHz audio nulls the carrier on the eighth carrier null for a 25 kHz deviation. The carrier must go through 8 nulls as the audio tone frequency is increased and the analyzer provides an ideal view as well as a very accurate frequency count of the sideband spacing.

The tone needs to be set very accurately in frequency. An independent frequency counter can be used at the audio signal source, or you can use the analyzer's two markers in frequency count mode to set the audio signal frequency. From Figure 136 the audio frequency is the spacing between adjacent sidebands.

Observation While Modulated

Modulators are not always on the bench, so it is handy to observe their deviation while on-line. A simple guideline called Carson's rule says that the occupied bandwidth of an FM signal is about 2 × (peak deviation + highest audio frequency). For television FM this is 2 × (25 kHz + 15 kHz), or 80 kHz. Use the analyzer to observe the modulation directly by tuning to an audio carrier in a span of 100 kHz with the other controls set automatically. To collect the amplitude data, select the trace maximum and let information accumulate. Rock and roll music accumulates faster than a talk show. Figure 137 shows such a trace. The deviation is only 24 kHz out of a possible 80 kHz.

TIP   **Use maximum trace hold to observe the FM bandwidth directly.**

Here is how to use the markers to estimate the occupancy bandwidth:

+ Peak the marker on the signal response.
+ Move the second marker to the side until it reaches about 6 dB down.
+ Move the original marker to the opposite side of the trace response.
+ Read the 6 dB bandwidth from the marker readout.

In the figure, the readout shows 23.8 kHz, well within the allocated audio bandwidth.

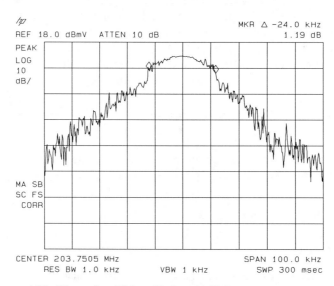

Figure 137. Observing FM audio bandwidth.

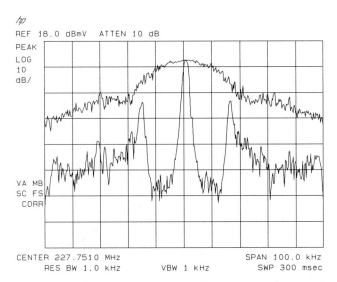

Figure 138. A 100 kHz span showing a channel audio with pilots and subcarriers.

Other signals in the audio modulation band can be monitored, too. Figure 138 shows the spectrum of an audio signal with pilot tones and subcarriers underneath a maximum hold of the signal's modulation. The pilot trace is captured by using the analyzer's single sweep until you catch a clear spectrum.

Slope Detection Method for Measuring FM

If the spectrum analyzer is fixed-tuned and the center frequency is set to either side of the audio signal, the analyzer's resolution bandwidth filter detects the frequency changes of the audio signal. In Figure 139 as the audio signal moves left and right along the slope of the analyzer's filter, the frequency shift is seen on the display as an amplitude change. If the slope of the filter is calibrated, peak deviation can be measured directly.

TIP

Use slope detection along with program audio to observe the FM deviation as it changes.

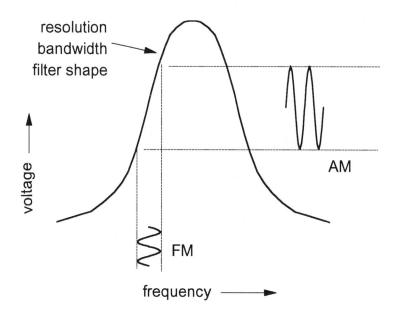

Figure 139. FM translation into amplitude swings by the analyzer's resolution filter.

Here is how to calibrate your analyzer for use in measuring the peak deviation of an audio signal:

* Using the calibrator signal of the analyzer, set the span to 400 kHz and the resolution bandwidth to 100 kHz.
* Display the calibration signal as in Figure 140 using two markers on the most linear portion of the resolution filter's side. Separate the markers by 50 kHz to represent the peak-to-peak deviation of the television audio.
* Record the amplitude difference between the markers, in this case 7.53 dB, and the distance from the calibration signal center to the analyzer's center frequency, 120 kHz. When the amplitude swing is 7.53 dB at an audio offset of 120 kHz, the peak-to-peak deviation is 50 kHz.
* Now connect the cable signal and make a measurement.

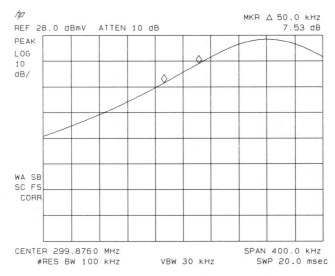

Figure 140. Calibrator signal used to determine the slope detection ratio for FM deviation measurement.

Example 51. Calculate the peak deviation using the slope detection method.

After calibrating your analyzer as above, center an audio carrier to be tested in the display with a span of 50 kHz. Change the resolution bandwidth to 100 kHz and put the analyzer into its zero span mode. Move the signal response to the top graticule.

Offset the center frequency 120 kHz higher or lower than its current setting. In this case, the center of the audio is at 59.750 MHz. The offset frequency is 59.750 − 0.120 = 59.630 MHz. The amplitude variations represent the changes in the audio carrier's frequency. From Figure 140 it was learned that 50 kHz is represented by 7.53 dB. Change the amplitude scale to 2 dB/division and take single sweeps until you see the widest swing. See Figure 141. Here the extreme amplitude shift is almost 8 dB, showing excess deviation. However, if you observe the display for a while with continuous sweep, the average is far less. So it appears that the channel is neither under- nor overmodulated. The peak-to-peak deviation is $(8/7.53) \times 50$ kHz = 53.1 kHz. Peak deviation is one half this, or ±26.6 kHz.

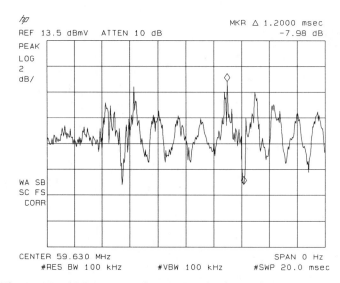

Figure 141. AM representing the audio signal FM.

Accuracy

FM deviation calibration using the Bessel null technique is accurate to within 1%, but the procedure only calibrates a single ±25 kHz deviation point. The procedure using the program material to carve out a trace maximum of the FM deviation is at best the accuracy of the span, or about ±3% of the full span. But this number is dependent upon program material. The slope detection technique yields 5% accuracy for the peak-to-peak deviation, if you assume that the program material had sufficient white signal to demonstrate the modulator's deviation.

Deviation measurements with program audio are only a quick look for preventive maintenance, not calibration measurements.

Digital Signal Levels

Although digital signals are far from the majority of carriers in cable television systems, they are gaining popularity as subscribers are made aware of their benefits. Your concern in delivering the promised service is to be able to set the levels of these digital signals uniformly with each other and with the analog carriers. This section shows you how to use the spectrum analyzer to make consistent measurements and recommends which of the analyzer's features to use.

How Digital Signals Look in Frequency Domain

A digital signal is usually a broad-band lump of noiselike signal. If you change the resolution bandwidth, a digital signal usually changes amplitude as does a noise signal, that is, about 10 dB of signal increase with every decade of resolution bandwidth increase. Different types of digital modulation may respond differently to changes in the analyzer settings.

Analyzer Tools to Help Quantify

Your job is to set the digital signal carriers consistent with the other analog carriers in the system as well as relative to the other digital signals. With the analog signals this is not difficult because either the modulation is a tops-down AM, as in the video signal, or an FM signal whose level does not change. In either case, maximum hold with a 300 kHz bandwidth does the job. When these setting are used with the digital signals of Figure 142 it takes a long time to get a smooth maximum response for amplitude comparisons. The lower trace is input, and the upper trace is the maximum hold trace.

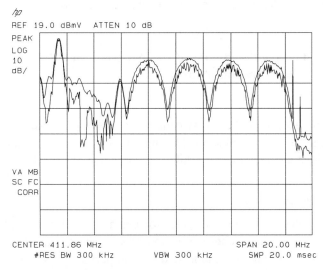

Figure 142. Digital signals with 300 kHz resolution and video bandwidths.

> **TIP** **Digital signals look like signals but respond like noise. They are neither!**

More important, is the maximum response proportional to the signal level? From observations in Chapter 7, the maximum response of noise is not related to the noise power. An average value is more valuable as a power index because digital signals appear as noise, but behave more like a packet of impulses that hit the analyzer's detector with random spikes. These impulses are such that narrow resolution and video bandwidths do not average the level enough to give you consistent amplitude readings.

> **TIP** **Most noise marker functions do not provide consistent level measurements on digital signals.**

The noise marker uses the sample detection mode and an average of points on each side of the marker. Observations show that even the noise marker has the amplitude jitters with narrow bandwidths. Averaging with a small video to resolution bandwidth ratio does average the digital signal levels, but takes a very long time. The best compromise between speed and averaging is the digital video averaging shown in Figure 143.

> **TIP** **Use digital video averaging with a count of 100 for consistent signal level measurements.**

The video average sampling is set to 100 sweeps, so the averaging takes a few seconds with the sweep time set to 20 ms full span. This measurement technique is responsive enough to let you adjust the signal levels while they are being observed.

Channel Power Measurements

Without going into the detail of their operation, some analyzers offer channel power and adjacent channel power functions. These are helpful for setting the absolute powers of similar digital signals, but are not recommended for comparing the power or bandwidths of digital and analog signals. Figure 144 shows an adjacent channel measurement.

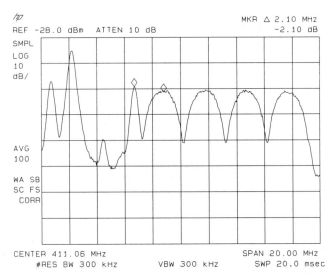

Figure 143. A digital signal level compared to an adjacent audio subcarrier using digital video averaging.

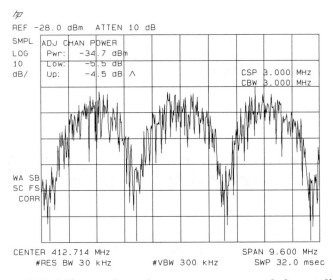

Figure 144. Adjacent channel power measurement helps set digital channel power levels.

One last tool for digital signal measurement is power bandwidth. Figure 145 shows a digital signal whose channel occupancy is 3 MHz measured for its 99% power bandwidth.

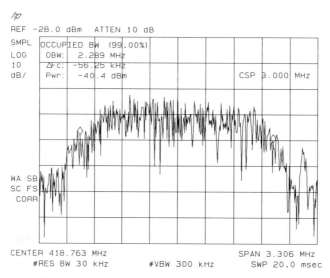

Figure 145. Power bandwidth for 99% of the power.

TIP

While channel power functions can give consistent digital signal power readings, they cannot be used for comparing analog and digital signals.

Conclusions

You are the best judge of what levels to set digital signals in your plant. Your spectrum analyzer has tools to make that job easy. For setting levels of digital signal.

◆ Use video averaging with a high average count setting, such as 100.
◆ Use 300 kHz resolution and video bandwidths.
◆ Use adjacent channel power and channel power functions for comparing digital signal levels.
◆ Be consistent in measurement procedures.

Depth of Modulation

Depth of modulation is a video signal quality measurement required for maintaining broadcast-television picture quality. Differential gain and differential phase suffer when depth of modulation is poor. Since new FCC regulations require differential phase and differential gain to be tested, depth of modulation can be a good first look at video quality. This section shows you how to observe depth of modulation on carriers with program material. Appendix F discusses the video measurements differential phase and gain, and CLDI.

Television Picture Modulation

Chapter 3 describes the makeup of the video carrier and how it is modulated. To understand what observations can be made with the spectrum analyzer look at a single horizontal line of video as shown in Figure 146.

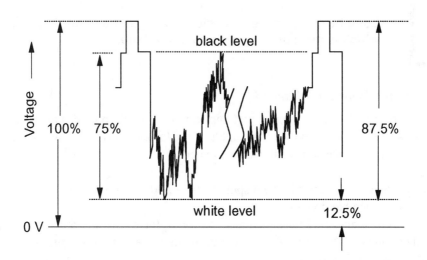

Figure 146. A single line of horizontal video.

Simply stated, the transmission standards set the limits of modulation depth for the black and white of a picture. The depth of modulation is measured as the percentage of the total amplitude change of the carrier, where the carrier is shown as 100% in the figure. Standard NTSC video modulation requires that the black/white range be contained in 87.5% of the carrier envelope, from full carrier to no carrier. At the sync tips the carrier is at maximum amplitude. When a white portion of the picture is encountered, the carrier amplitude is reduced to 12.5% of the carrier maximum.

> **TIP**
>
> **The spectrum analyzer can measure the depth of modulation on live channels.**

When Depth of Modulation Goes Bad

Low contrast is indicative of undermodulation. The picture appears as faded photographs. Overmodulation, that is, exceeding the 87.5% modulation, causes the video carrier to disappear during the peak white video. When the carrier disappears, the television receiver searches for a signal to process at the television field rate, which is heard as a loud buzzing. This distortion is called sync buzz.

> **TIP**
>
> **Loud buzzing on the television is a sign of overmodulation, which is measured by depth of modulation.**

Observing with Program Material

The measurement technique is simple; just duplicate Figure 146 on the analyzer display, and use markers to measure the relative depths of the lowest, or deepest, modulation swings. To set up the analyzer for this measurement:

- ◆ Center the visual carrier in the display using a 300 kHz resolution bandwidth and a 300 kHz video bandwidth.
- ◆ Select zero span, bring the signal near the top of the display, and change the vertical scale to linear, or voltage. The bottom graticule represents zero volts.
- ◆ Set the sweep time for 5 μs.
- ◆ Adjust the amplitude to get the maximum responses to the top graticule. This is the 100% carrier level.
- ◆ If your analyzer is so equipped, allow the collection of minimum trace responses in one trace while another trace collects the maximum.
- ◆ Observe the depth directly, or use markers to measure the relative level of the minimum swing.

Figure 147 shows the results. In this analyzer, the display is conveniently divided into eight vertical divisions, each representing 12.5% voltage. Any swing into the lowest division is overmodulation.

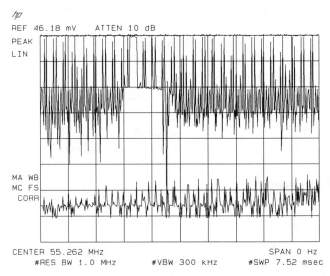

Figure 147. Depth of modulation on program material.

On an analyzer display with eight linear divisions, overmodulation sends signal responses down to the lowest division, over 87.5%.

Depth of Modulation Accuracy

At best, the accuracy is ±3%, the uncertainty of the analyzer's linear scale fidelity. Using program material makes the measurement much more subjective than would the vertical interval test signal, VITS, or video test signal generator.

Summary

This chapter shows you how to make those audio and video measurements that are important but not critical to compliance. Depth of modulation, FM deviation, and digital carrier amplitude can all evoke subscriber complaints, but no direct FCC rules are broken.

The spectrum analyzer is a good observation tool for these measurements because, what it lacks in accuracy, it makes up for in speed and convenience.

- Use the analyzer and a signal generator to calibrate FM peak deviation on a modulator.
- Use the analyzer alone to observe FM deviation in two ways: direct observation of the signal bandwidth, or, more accurately with slope detection, converting the analyzer's resolution filter to an FM detector.
- Use digital video averaging and consistent analyzer settings to set digital signal levels.
- Set the analyzer to observe depth of modulation on program material with just a few keystrokes and observe extreme over- or undermodulation.

Selected Bibliography

- Bowick, Chris. "A Review of Frequency Modulation", *Communications Engineering & Design,* July 1988.
- Benson, K. Blair, and Whitaker, Jerry. *Television Engineering Handbook.* rev. ed., McGraw-Hill, Inc., 1992.
- Edgington, Francis M. Hewlett-Packard Company, Microwave Instruments Division, Santa Rosa, CA, *personal communication*, September 1994.
- Hranac, Ron and Johnson, Steve. *How to Adjust Audio Carrier Deviation.* Communications Technology Publications Corp., November 1989.

Appendix A - Tables for Unit Conversion

Conversion Among Different Units

Often you will have to convert one unit of measure to another, especially when dealing with correction factors.

Table 10. Formulas for converting from one unit to the other, assuming the same impedance.

		TO				
		volts, V	watts, W	dBm	dBmV	dBμV
FROM	volts, V	V	V^2/Z	$10 \log \frac{V^2}{10^{-3} \times Z}$	$20 \log \frac{V}{10^{-3}}$	$20 \log \frac{V}{10^{-6}}$
	watts, W	$\sqrt{W \times Z}$	W	$10 \log \frac{W}{10^{-3}}$	$20 \log \frac{\sqrt{W \times Z}}{10^{-3}}$	$20 \log \frac{\sqrt{W \times Z}}{10^{-6}}$
	dBm	$\sqrt{\log^{-1}\left(\frac{dBm}{10}\right) \times Z \times 10^{-3}}$	$\log^{-1}\left(\frac{dBm}{10}\right) \times 10^{-3}$	dBm	$dBm + 30 + 20 \log (\sqrt{Z})$ (+48.75 for 75 Ω)	$dBm + 90 + 20 \log (\sqrt{Z})$
	dBmV	$\log^{-1}\left(\frac{dBmV}{20}\right) \times 10^{-6}$	$\left[\log^{-1}\left(\frac{dBmV}{20}\right)\right]^2 \times \frac{10^{-6}}{Z}$	$dBmV - 30 - 20 \log (\sqrt{Z})$	$dBmV$	$dBmV + 60$
	dBμV	$\log^{-1}\left(\frac{dB\mu V}{20}\right) \times 10^{-6}$	$\left[\log^{-1}\left(\frac{dB\mu V}{20}\right)\right]^2 \times \frac{10^{-12}}{Z}$	$dB\mu V - 90 - 20 \log (\sqrt{Z})$	$dB\mu V - 60$	$dB\mu V$

Example 52. Converting from voltage to dBmV.

Your spectrum analyzer reading for field strength is +14.2 dBmV/meter. The specification is required in dBμV/meter so you need to convert the reading. Select the correct formula across the dBmV FROM row down from the dBμV TO column. The formula from Table 10 is dBμV + 60. The answer is +14.2 dBmV + 60 = +74.2 dBμV. Note that the conversion is independent of impedance because both units were referenced to volts.

Example 53. Converting from watts to dBmV.

A preamplifier input is specified for an input less than 3×10^{-5} Watts for its optimum distortion performance. The preamplifier impedance is 75 Ω. What is this limit in dBmV?

Find the formula on the row labeled watts, W, and down from the column labeled dBmV. It is $20 \log ((W \times Z)^{1/2}/10^{-3})$ where $(...)^{1/2}$ is the same as square root, $\sqrt{}$. Then $20 \log [(3 \times 10^{-5} \times 75)^{1/2}/10^{-3}] = 20 \log (47.4) = +33.5$ dBmV in 75 Ω

Converting dbmv with Different Impedances

A common conversion is from dBmV in one impedance to dBmV in another impedance.

Table 11. Conversion of dBmV for different impedances. Add the value in the table to the dBmV to be converted FROM.

		TO dBmV in impedance Z_2				
		50 Ω	75 Ω	300 Ω	600 Ω	Z_2 in Ω
FROM dBmV in impedance Z_1	50 Ω	0	+1.76 dB	+7.78 dB	+10.79 dB	$+10\log_{10}\dfrac{Z_2}{50}$
	75Ω	-1.76 dB	0	+6.02 dB	+9.03 dB	$+10\log_{10}\dfrac{Z_2}{75}$
	300 Ω	-7.78 dB	-6.02 dB	0	+3.10 dB	$+10\log_{10}\dfrac{Z_2}{300}$
	600 Ω	-10.79 dB	-9.03 dB	-3.01 dB	0	$+10\log_{10}\dfrac{Z_2}{600}$
	Z_1 in Ω	$+10\log_{10}\dfrac{50}{Z_1}$	$+10\log_{10}\dfrac{75}{Z_1}$	$+10\log_{10}\dfrac{300}{Z_1}$	$+10\log_{10}\dfrac{600}{Z_1}$	$+10\log_{10}\dfrac{Z_2}{Z_1}$

Example 54. Convert dBmV from 50 to 75 Ω.

Power of a visual carrier according to a 50 Ω spectrum analyzer is −3.6 dBmV. Disregarding the mismatch error, what is the power in 75 Ω?

On the FROM side select the 50 Ω row, and read the conversion down from the 75 Ω column. The conversion is +1.76 dB. The power in 75 Ω is −3.6 dBmV + 1.76 dB = −1.84 dBmV.

Appendix B - Spectrum Analyzer Block Diagram and Operation Topics

Relating the Block Diagram to Front Panel Controls

This block diagram makes the connection between the analyzer's internal circuitry and the controls you see on the front panel. The word **couple** joining some controls and analyzer functions indicates that a change of one control affects the other. The simplest example is the relation between frequency span and resolution bandwidth. Under coupled conditions, the bandwidth will increase as a percent of the frequency span.

Figure 148 on the following page shows a spectrum analyzer block diagram and indicates how the front panel controls relate to the analyzer circuits.

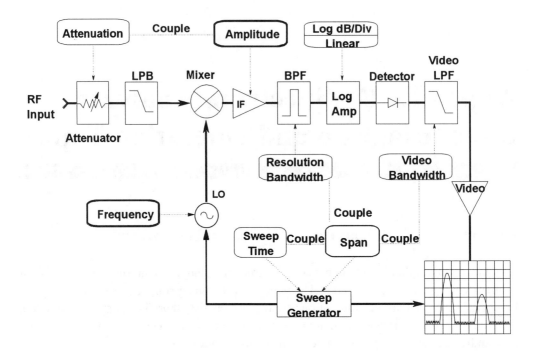

**Figure 148. Block diagram of spectrum analyzer relating front
panel controls to analyzer circuits.**

Impedance Matching the Spectrum Analyzer Input

The general-purpose spectrum analyzer, and most microwave analyzers, have a 50 Ω input impedance. If you have such an analyzer sitting on the shelf because it does not conform to the 75 Ω impedance of your system, use one of the matching networks suggested in Figure 149 and put it to work.

Matching circuit	Circuit diagram	Voltage ratio V_{50}/V_{75}	Power ratio in dB $10 \log [(V_{50}^2/50)/(V_{75}^2/75)]$ $=10 \log (3/2)(V_{50}/V_{75})^2$	Insertion loss
Direct connection of cables	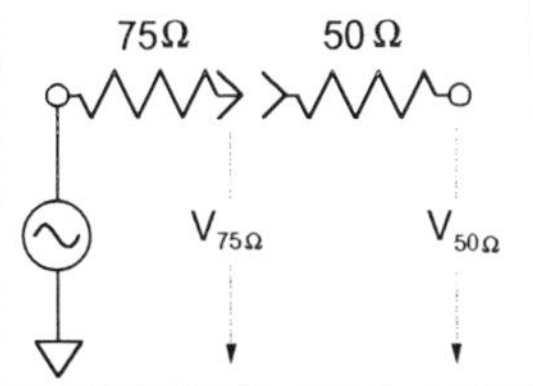 75Ω 50 Ω $V_{75Ω}$ $V_{50Ω}$	$V_{50} = V_{75}$	Who knows?	None except when connectors break
Matching transformer	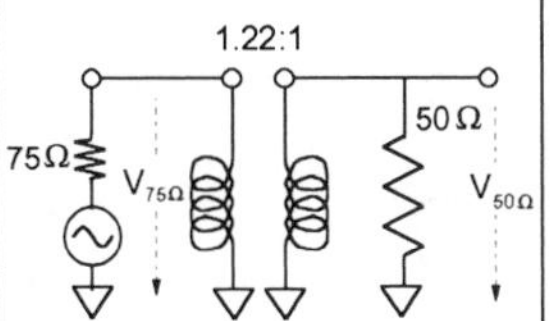 1.22:1 50 Ω 75Ω $V_{75Ω}$ $V_{50Ω}$	0.82	$10 \log [(3/2)(0.82)^2]$ $= 0$ dB	0.75 dB, often specified
Minimum loss pad	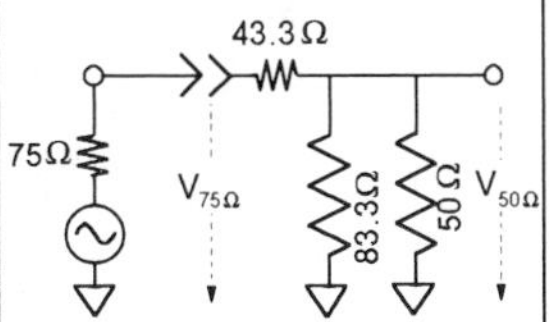 43.3 Ω 75Ω $V_{75Ω}$ 83.3 Ω 50 Ω $V_{50Ω}$	$(75-43.3)/75$ $= 0.423$	$10 \log [(3/2)(0.423)^2]$ $= -5.71$ dB	5.7 dB usually specified
Matching resistor	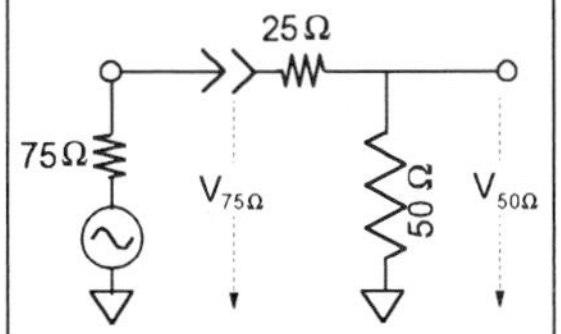 25 Ω 75Ω $V_{75Ω}$ 50 Ω $V_{50Ω}$	$[50/(25+50)]$ $= 0.667$	$10 \log [(3/2)(0.667)^2]$ $= -1.76$ dB	1.76 dB

Figure 149. Selection of matching networks available along with thin circuits, benefits and problems.

The direct connection of cables is a poor match and is not recommended especially since connector damage is common. A matching transformer gives the best performance, but frequency range is limited. VSWR is approximately 1.2. The overall best solution for 50 Ω spectrum anlayzers is a minimum loss pad which has >20 dB return loss (very good VSWR). For a simple match when no others are available, you can use a matching resistor.

Appendix C - Spectrum Analyzer Accuracy Topics

Accuracy Information

The primary source of spectrum analyzer accuracy information is its data sheet. The examples in this book that compute the analyzer's accuracy use information from the following example data sheet. The use of the data sheet can be confusing because of the number of specifications and conditions. To help you make accuracy computations, two work sheets are included to guide you through the frequency and amplitude accuracy process for this sample data sheet.

Sample Data Sheet

Frequency Specifications

Frequency Range: 1 MHz to 1.8 GHz

Frequency Reference Error

Aging: $\pm 1 \times 10^{-7}$/year

Settablity: $\pm 2.2 \times 10^{-8}$

Temperature stability: $\pm 1 \times 10^{-8}$

Frequency Accuracy

Frequency span ≤10 MHz ±(frequency readout × frequency reference error + 3.0 % of span + 20% of RBW + 100 Hz)

Frequency span >10 MHz ±(frequency readout × frequency reference error + 3.0 % of span + 20% of RBW)

Marker Count Accuracy (S/N ≥25 dB, RBW/span ≥ 0.01)

Frequency span ≤10 MHz ±(marker frequency × frequency reference error + counter resolution + 100 Hz)

Frequency span >10 MHz ±(marker frequency x frequency reference error + counter resolution + 1 kHz) where counter resolution selectable from 10 Hz to 100 kHz

Frequency Span

Range 0 Hz (zero span), 1 MHz to 1.8 GHz

Accuracy ±2% of span, span ≤10 MHz; ±3% of span, span >10 MHz

Amplitude Specifications

Amplitude Range: Displayed average noise level to +72 dBmV

Maximum Safe Input

Peak power +72 dBmV (0.2 watts), input attenuation ≥10 dB

Gain compression >10 MHz ≤0.5 dB (+39 dBmV at input mixer)

Reference Level

Range: Same as Amplitude Range

Resolution: 0.01 dB for log scale, 0.12 % of ref level for linear scale

Accuracy (Referred to +29 dBmV ref level): +49 to −10.9 dBmV, ±(0.3 dB + 0.01 × dB from + 29 dBmV)

Frequency Response

Absolute ±1.5 dB

Relative flatness ±1.0 dB

Calibrator Output

Frequency 300 MHz ±(300 MHz × frequency reference error)

Amplitude +28 dBmV ± 0.4 dB

Input Attenuator

Range: 0 to 70 dB in 10 dB steps

Accuracy: 0 to 60 dB ±0.5 dB at 50 MHz, referenced to 10 dB attenuation setting, 70 dB ±1.2 dB at 50 MHz, referenced to 10 dB attenuation setting

Display Scale Fidelity

Log incremental accuracy: ±0.2 dB/2 dB, 0 to 70 dB from reference level

Log max cumulative: ±0.75 dB, 0 to −60 dB from reference level, ±1.0 dB, 0 to −70 dB from reference level

Linear accuracy: ±3% of reference level

Internal Preamplifier

Frequency range 1 MHz to 1.0 GHz

Gain ≥20 dB

Noise figure ≤5 dB

Accuracy Computation Worksheets

Frequency

Models	Normal Marker Uncertainty ± Frequency		Counter Marker ± Frequency > 25 dB S/N	
	Single	Delta	Single	Delta
Frequency accuracy options	Freq × Freq Ref + 3% Span +20 % Res BW +100 Hz(span< 10MHz)	Not specified. Typically ± 3% of span.	Freq × Freq Ref + Counter Resolution + 100 Hz × N	Not specified. Typically = Delta Freq × Freq Ref + 2 x counter res + 200 Hz × N
No frequency accuracy options	N × 5 + 2% Span	Not specified. Typically ± 5% of span.	Not applicable	Not applicable

Where N for the analyzer frequency is:

N	Nominal Frequency Range
1	< 6.5 GHz
2	6 to 12 GHz
3	12 to 19 GHz
4	19.1 to 22 GHz
4	19.1 to 26.5 GHz

Amplitude

CONSIDERATIONS		UNCERTAINTIES	± dB
Are the signal responses smooth and stable?	Yes	Go to the next step.	
	No	Accurate amplitude measurement may not be possible; see < section on display, detection, stability>.	
Has the spectrum analyzer been calibrated with internal calibration signal and routines?	Yes	Go to the next step.	
	No	Perform calibration routines listed in the Operating Manuals, then continue.	
MEASURE THE SIGNAL. If making a relative measurement, measure the higher signal first. Use the marker function. Let the spectrum analyzer select the bandwidths, sweep time, and input attenuator settings.	Amplitude: Reference level: Frequency: Frequency range (band): Resolution bandwidth: Input attenuator:		
Is the signal peak higher than the lowest display division in 10 dB/division scale?	Yes	Go to the next step.	
	No	Decrease the reference level until signal is higher than the lowest display division.	
Is the signal peak more than 5 dB higher than the displayed average noise? An amplitude error makes the signal appear higher than actual.	Yes	Go to the next step.	
	No	Increase the displayed signal to noise by decreasing the resolution bandwidth. Narrow frequency span if practical.	
Is the measurement between two signals (RELative), or of one signal (ABSolute)?	Rel	Go to the next step.	
	Abs	Go to ABSOLUTE AMPLITUDE MEASUREMENT.	

Continued

MEASURE THE SECOND SIGNAL. If possible, let the spectrum analyzer select the bandwidths, sweep time, and input attenuator settings.	Amplitude: Reference level: Frequency: Frequency range (band): Resolution bandwidth: Input attenuator:		
Is the signal level of either signal +28 dBmV?	No	Record reference level accuracy as double the amount on the data sheet.	
	Yes	Record reference level accuracy.	
Was the resolution bandwidth changed between measuring the first and second signal?	No	Go to the next step.	
	Yes	Record the resolution bandwidth switching uncertainty.	
Are the signals in different harmonic bands?	No	Go to the next step.	
	Yes	Record the highest relative frequency response of the two bands regardless of attenuator setting. Add the band switching uncertainty.	
Are the signals in the same band separated by more than 5% of the band range?	No	Go to the next step.	
	Yes	Record the frequency response for the band regardless of the setting.	
Was the input attenuator set to 10 dB during the measurement of EITHER, NEITHER, or BOTH of the signals?	Either	Record the input attenuator switching uncertainty plus the attenuator repeatability.	
	Neithe r	Double the input attenuator switching uncertainty plus double the attenuator repeatability.	
	Both	Go to COMPUTE ACCURACY.	
ABSOLUTE AMPLITUDE MEASUREMENT: Record the frequency response uncertainty referenced to the CAL signal, plus the CALibrator accuracy.			
Is the input attenuator in the 10 dB setting?	No	Record input attenuator switching uncertainty.	
	Yes	Go to the next step.	
COMPUTE ACCURACY		**TOTAL UNCERTAINTIES:**	

Appendix D - Carrier-to-Noise Measurement Topics

Overview

This appendix is a supplement to the C/N measurement of Chapter 7. It covers:

- Spectrum analyzer noise smoothing
- Noise power density computation
- Details on noise correction procedures including the noise power equivalent bandwidth
- Preamplifier gain measurement

Smoothing the Random Amplitude of Noise Variations

The spectrum analyzer displays noise as a ragged line across the lower portion of the display. Each point along the baseline represents a single noise power density reading of the noise passing through the resolution bandwidth filter as the analyzer sweeps across the span. The amplitude variations are caused by the analyzer's detection of the random amplitude variations of the noise. What the illustration cannot show is what happens over time. At any frequency point the average of the noise power over time produces a value that represents the noise power at that point. Since the noise at adjacent frequency points usually average to similar levels, a flat response over frequency is displayed.

A noise power average is done with the analyzer's low pass filter, called the video filter, which is shown in Figure 150, a simplified diagram of the analyzer's IF.

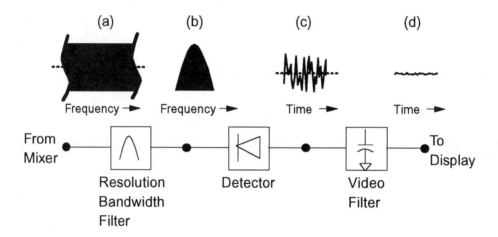

Figure 150. The video filter effect on noise in the analyzer's IF.

The broadband noise from the analyzer's mixer enters the resolution filter at (a). The resolution filter cuts the total noise power down to the IF resolution band shape in (b), but retains the relatively wide bandwidth of the noise. The detector, represented by the diode in the figure, produces an envelope of the noise signal. This noise voltage still has a bandwidth equal to the resolution bandwidth, allowing the noise energy to produce large voltage swings, as seen in (c). Averaging these voltage swings to a single point, which represents the noise level at the tuned frequency of the analyzer, requires the video filter to be narrowed in bandwidth much narrower than the resolution bandwidth setting. The video filter operates as a capacitor across the signal, damping out fast amplitude variations and charging to an average noise value, as in (d).

Since adjacent frequency points are averaged to similar levels, the effect is to produce a single line across the analyzer's display which represents noise power density over frequency.

Displayed Noise Level Dependent upon the Resolution Bandwidth

Noise levels detected and displayed are dependent upon the resolution bandwidth used. The wider the bandwidth, the higher the displayed noise. Figure 151 illustrates. In this figure, successive sweeps are taken, each with a ten times higher resolution bandwidth. Video filtering is used to smooth the noise. The top trace is a resolution bandwidth of 300 kHz, the middle for a bandwidth of 30 kHz, and the bottom for a bandwidth of 3 kHz. Each has about 10 dB less power than the next, yet each represents the same noise power density. The levels are different because the different resolution bandwidths allow different levels of noise power to the detector. The input noise level has not changed. As the bandwidth is increased, the level goes up.

The rule of thumb for the relationship between IF noise equivalent and resolution bandwidth is simple: Noise power increases 10 dB for every ten times increase of the resolution bandwidth. This rule is borne out in Figure 151, where the resolution bandwidths are changed by powers of ten, from 30 kHz at the top trace, to 3 kHz in the middle trace, to 300 Hz on the bottom trace. Noise power changes are expressed in the following equation.

$$\text{Noise power change in dB} = 10 \log_{10}(\text{RBW}_1/\text{RBW}_2)$$
$$\text{Equation 5}$$

where the RBW terms are the two measurement resolution bandwidths.

Example 55. Noise level changes with resolution bandwidth.

The noise power reads −52 dBmV with the resolution bandwidth set at 300 kHz. Calculate the noise reading when the resolution is changed to 10 kHz.

The change in power level is $10 \times \log_{10}(10 \text{ kHz}/300 \text{ kHz}) = -14.8$ dB. The minus sign says that the power level is lower on the display, the 300 kHz bandwidth being the reference bandwidth. It should become natural to predict which way the noise moves when changing the resolution bandwidth because the level changes with the width of the bandwidth.

TIP

For every ten-time change in resolution bandwidth, expect a 10 dB change in noise level. The wider the bandwidth, the higher the noise.

Noise Power Density

Calculating Noise Power Density

The noise power density is the noise power level read on the spectrum analyzer, corrected by the resolution bandwidth ratio. As specified in the FCC regulations for C/N, the noise power density needs to be referenced to 4 MHz. Equation 5 shows the relationship, an the following example shows a calculation.

$$\text{Noise power density} = \text{noise power reading} \times 10 \, \log_{10}(\text{RBW}_{\text{REF}}/\text{RBW}_{\text{ANALYZER}})$$
Equation 6

where $\text{RBW}_{\text{ANALYZER}}$ is the resolution bandwidth used in the measurement

RBW_{REF} is the noise power density reference bandwidth

Example 56. Noise power density.

Determine the noise power density of the noise level shown in Figure 151.

A marker placed on the center noise display reads −61.08 dBmV. The resolution bandwidth for this trace was 30 kHz. The noise power density is given by Equation 5. Noise power density = −61.08 dBmV + 10 $\log_{10}$(4 MHz/30 kHz) = −61.08 dBmV +21.25 dB = −39.83 dBmV/4 MHz.

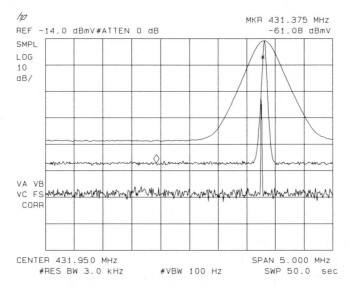

Figure 151. The same noise input measured with different resolution bandwidths.

Remember:

- ◆ The changes are in dB and are added and subtracted, not multiplied and divided.
- ◆ The result is in dBmV per reference bandwidth frequency, 4 MHz in this case.

Noise power density must have a stated or implied reference bandwidth when it is written, otherwise the level is meaningless. For convenience, Table 11 gives the corrections to noise power readings for the most-used spectrum analyzer resolution bandwidths. Add the correction to the analyzer reading of noise to get the noise power density value for C/N calculation.

Table 11. Noise power density corrections.

Spectrum analyzer resolution bandwidth	4 MHz Correction dB
10 kHz	26.02
30 kHz	21.25
100 kHz	16.02
300 kHz	11.25

Example 57. Correcting noise power for 4 MHz bandwidth.

Correct the power level in Figure 151 to a 4 MHz resolution bandwidth.

The noise reads −61.08 dBmV on the analyzer in a 30 kHz bandwidth. From Table 11, the 30 kHz correction is +21.25 dB. The noise power density for a 4 MHz bandwidth is −61.08 + 21.25 dB, or −39.83 dBmV/4 MHz.

Sampling Detection Used for Noise

The spectrum analyzer is set up to make CW signal measurements when you turn it on. One of the settings is the detector type. The detector, referring again to Figure 150, is the diode in the IF circuit that converts the filtered IF signal of every swept frequency point to a signal amplitude value to be displayed as a trace point. For CW signals, whose maximum response represents its RMS power, the detector is set to measure the peak of each frequency sample during the sweep. Since noise signal power is not a peak measurement, but rather an average power density, each frequency point must be averaged from a random sample of the entire range of amplitudes at each frequency point.

TIP   **Use sample detection for measuring noise, and peak detection for measuring carrier levels.**

Figure 152 illustrates the difference responses from a peak and a sample detection mode. The noise signal in (a) is shown with random amplitude variations during each frequency interval of the analyzer's sweep.It is the detector's job to condense all the variations in one interval to a single point to be displayed. If the analyzer is set to its default peak detection mode, as if it were going to measure CW, or a modulated CW signal, like the visual carrier, the noise signal would be displayed as in (b). The dashed line at (b) represents a false average of the noise; it is higher than the average. In order to measure the average of all the amplitude variations at each point, the detector needs to be set to sample the noise randomly. This detector mode is called the sample. It samples by arbitrarily taking a reading periodically, not a maximum as with the peak detection mode. The results of this

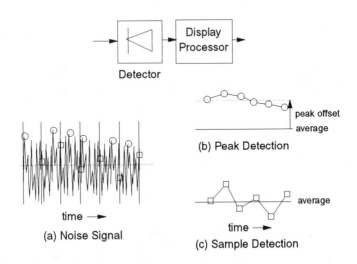

Figure 152. Sample and peak detection at a single frequency point.

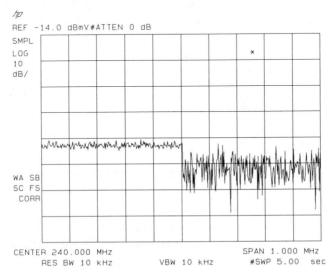

Figure 153. Noise viewed with peak (left side) and sample detection (right side).

sample detection and its average are represented by (c). With noise, the average at each interval is desired, so the detector is set to sample a single value in each interval, as shown in (c). In this case the sample is taken at the end of each interval to assure that the points collected are random samplings of the noise in a sort of uniform randomness.

Figure 153 dramatically shows the different levels when using peak detector (on the right half of the display) and the sample detector (on the left half of the display). Note that the maximum peaks are rarely captured with the sampled detection mode. This is because they do not occur often enough to be caught in the periodic sampling used during sample detection. But the peak detection mode saves the maximum response in every interval no matter how seldom high responses occur.

Conclusion: Always use sample detection when making noise power measurements.

Terminate the Analyzer Input When Measuring Its Noise

Note that there is a response on the right side of Figure 154, which is the analyzer noise with no input. The source of this signal could be an internal residual signal of the analyzer itself, or, more likely, a radiated signal at the open input port of the analyzer. It is good practice to terminate the spectrum analyzer input with a load whose resistance is the same as its input impedance when making analyzer noise measurements. The load assures that no stray radiated signals or broadband noise is affecting the readout. Radiated broadband or impulse noise, such as generated by power generators, automobile ignitions, and computers, is eliminated by shields inside most modern spectrum analyzers. But external hardware, such as open connectors, poorly shielded input cables, preamplifiers, or filters can act as antennas for noise and spurious CW signals which can cause inaccuracies. When in doubt, terminate the analyzer's input.

Measuring the Preamplifier Gain and Noise Figure

The gain of the preamplifier can be measured easily with the spectrum analyzer. The noise figure is possible to measure, but beyond the scope of this text. You must trust the manufacturer to provide accurate noise figure values and accuracy with the product. Often these parameters are written on the individual preamplifiers, indicating that they have been tested and documented during the manufacturer's final test. The preamplifier documentation should show the accuracy for gain and noise figure in terms of a ± dB value.

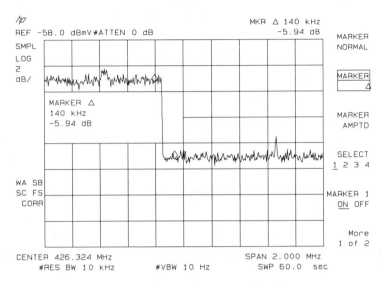

Figure 154. The disconnect test performed with a bit more care. Note the composite distortion signal on the left.

TIP   **Get the preamplifier noise figure from its manufacturer, but test the gain yourself.**

Here is how to measure the gain of the preamplifier.

Example 58. Measure preamplifier gain.

To measure the preamplifier gain simply insert the preamplifier between a known signal and the spectrum analyzer, and measure the signal increase in dB. Place a high visual carrier in the center of the 6 MHz display, and set the resolution and video bandwidths to 300 kHz. Bring the signal near the top of the display with the reference level control, keeping the attenuator at the value set when checking overload. If a tunable bandpass filter is used, tune the filter to maximize the carrier signal and readjust the signal amplitude on the analyzer.

Use the marker peak and marker to reference level functions to set the signal maximum on the reference level. If your analyzer does not have these functions, set the amplitude scale to 2 dB per division, and use the reference level control to bring the signal up to the top graticule. Record the signal level.

Remove the preamplifier from the circuit and adjust the reference level to bring the signal to the reference level again. Record the level and subtract from the amplified level to get the gain of the preamplifier at this frequency.

It is important to remove the preamplifier completely from the circuit, rather than just to turn off its power supply. A preamplifier without power attenuates the signal through it, giving a false reading of the carrier.

It is good practice to evaluate the preamplifier gain at three frequencies across the system to measure its frequency response. If wide variations of gain are seen, more carriers should be measured to confirm the preamplifier's frequency response and gain. If the gain is out of specification, the preamplifier may be damaged. Have it tested to factory specifications.

Measuring Filter IF Noise Equivalent Power Bandwidth, NEPBW

The trick is to measure, very precisely, the area under the Gaussian filter curve. The way this is done is shown in Figure 155. Each point on the analyzer's trace represents a power level of a segment of the span. This is represented in the figure by the vertical bars. Each vertical bar's width, in Hz, is the frequency span divided by the number of points across the display. The height is the power of that bar in watts. Because it is necessary to display the filter in dB to see its wide amplitude range, the power is read in dBmV and converted into watts.

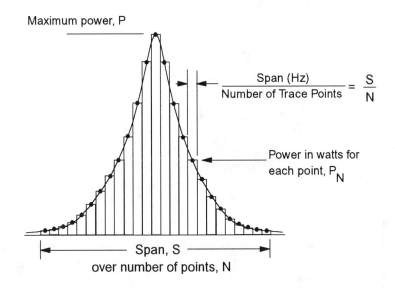

Figure 155. Measurement of the NEPBW of a Gaussian filter.

The total area under the filter curve is the summation of all the vertical bar areas, in units of watts × Hz, or watts – Hz. Each bar is its power in watts times the bar width in hertz. The width of each bar is the span divided by the number of points that make up a trace, or using the terminology in the figure, $P_N \times S/(N-1)$. The number of bars is one less than the number of points since one end point does not have a bar associated with it. The total area under the curve is the sum of all these bar areas, written:

$$P_1 \times S/(N-1) + P_2 \times S/(N-1) + \ldots P_N \times S/(N-1)$$

Since the term $S/(N-1)$ is common to all the terms, the area in watts – Hz becomes $S/(N-1) \times (P_1 + P_2 + \ldots P_N)$. The NEPBW is this total power divided by the maximum power response of the filter, P.

Using $N-1$ for the number of bars is precision overkill by more than 200 points, but becomes significant as the number of points approaches 100 or less.

The low values of P_N, the power at each trace point, may not contribute significant power. Values more than 40 dB down are discarded because they add only 0.0001 P. However, the values between 25 and 40 dB below P contribute measurable power.

$$NEPBW = (S/(N - 1) \times \Sigma P_N)/P$$
Equation 7

where NEPBW = IF noise equivalent power bandwidth in hertz
ΣP_N = sum of all the powers along the trace
P = highest response power in the trace
N = number of points in the trace
N − 1 = number of bars

TIP | **Measuring and calculating equivalent noise power bandwidth for your analyzer may require an instrument controller and computer.**

Calculating NEPBW without a computer is not easy. So if you need to do this test, contact the analyzer manufacturer or your associates who have access to a similar spectrum analyzer for trace data or results. If your computer and spectrum analyzer are equipped with an instrument interface and suitable software, and if the analyzer can be coaxed into sending its trace data to a computer file, then the following example shows how to compute NEPBW for your spectrum analyzer.

Example 59. Measuring NEPBW on a specific filter.

Measure the NEPBW on the spectrum analyzer's 30 kHz resolution bandwidth.

Since the analyzer filter shape follows any stable CW signal, its calibration signal or LO feedthrough response is used. Center the signal in the display in a 30 kHz resolution bandwidth. Set the video bandwidth to 30 kHz or wider. Change the detector to the sample mode. Set the span to five times the 30 kHz bandwidth so that the filter response at the start and stop frequencies are between 25 and 40 dB down from the peak response of the analyzer. Figure 156 shows the filter response using the

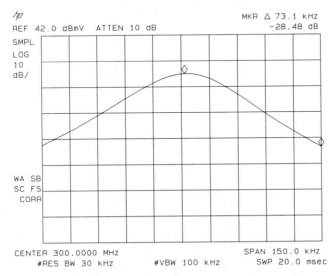

Figure 156.   The filter response for computing NEPBW.

300 MHz calibrator signal. The markers in the figure confirm that the end points are about 28 dB down.

Instruct the computer to send the trace data to a file. The values should be in dBmV. The display in the figure begins with values on the left of "−4.49, −4.30, −4.16, ..." dBmV, where the comma designates the end of one value. The center of the trace reads "...26.69, 26.69, ...26.74, 26.74, ...26.69..." where the value 26.74 dBmV represents the maximum peak, P, in dBmV, or 6.29E−03 watts. Use a computer spreadsheet or program to convert all the values to milliwatts, find the largest value, P, and the sum of the powers, ΣP_N. Table 12 shows sample values and computations from the trace data.

Now Equation 7 gives the NEPBW, (150,000/(401 − 1)*.566)/6.29E − 03 = 33,744 Hz. A straight-sided filter whose width is 33.74 kHz would allow the same noise amplitude measurement as the 30 kHz resolution bandwidth filter.

Trace point power, dBmV	Converted to watts
The first four left trace points	
-4.49	4.74E-06
-4.30	4.95E-06
-4.16	5.12E-06
-4.02	5.28E-06
*	
200 trace points	
*	
26.69	6.22E-03
26.69	6.22E-03
26.72	6.27E-03
26.74	6.29E-03
26.74	6.29E-03
26.74	6.29E-03
26.69	6.22E-03
26.69	6.22E-03
26.69	6.22E-03
*	
180 trace points	
*	
-3.81	5.55E-06
-4.05	5.25E-06
-4.19	5.08E-06
-4.35	4.90E-06
-4.46	4.77E-06
Total power in watts	0.57

Table 12. Sample calculations of NEPBW from the trace data in Figure 156. Note that the total power is based on 401 trace point. Only a few representative values are shown.

Once the NEPBW in Hz is found, a single correction can be calculated, in dB, for the noise level. The ratio of the two filter bandwidths is converted to dB by the following equation:

Noise Power Error Correction, $E_{ENP} = 10 \times \log$ (NEPBW/RBW)
Equation 8

where RBW is the analyzer's resolution bandwidth 3 dB.

For Example 59, the correction is $E_{ENP} = 10 \times \log (33.74/30) = 0.51$ dB. Does this value get added or subtracted from the noise power actually read on the analyzer? This can be confusing. The noise on the analyzer is read with a noise power bandwidth of 33.74 kHz, the value just calculated for this specific analyzer and this bandwidth setting. But the spectrum analyzer says it is using a 30 kHz bandwidth, so convert the noise power reading to the resolution bandwidth. Since the resolution bandwidth is narrower than the noise power bandwidth, the correction is subtracted from the analyzer reading; that is, the noise read on the analyzer is higher than would be read with a true 30 kHz square-sided noise filter.

Example 60. Correct noise reading for NEPBW.

The spectrum analyzer reads noise power at −62.58 dBmV when the resolution bandwidth is set to 30 kHz. What is the noise power corrected for NEPBW in the last example? What is the noise power for 4 MHz bandwidth?

With an IF noise equivalent power bandwidth of 33.74 kHz, the noise power is −62.58 − 0.51 = −63.09 dBmV/30 kHz. The noise power density could also be written −62.58 dBmV/33.74 kHz, but the odd value of bandwidth makes a complicated measurement even more confusing.

Either way, conversion of this noise to the required 4 MHz bandwidth ends up to the same value. The conversion of the −63.09 dBmV/30 kHz to 4 MHz is 10 log(4 MHz/30 kHz) = +21.25 dB. The noise power is −63.09 + 21.25 = −41.84 dBmV/4 MHz The conversion of the −62.58 dBmV/33.74 kHz is calculated as 10 log(4 MHz/33.74 kHz) = +20.74 dB. The noise power is −62.58 + 20.74 = −41.84 dBmV/4MHz.

Summary of Noise Measurement Procedures

The corrections and practices for making accurate and repeatable system noise measurements condense down to the following list:

- Use the good measurement practices.
- Correct for reference of noise power density from the resolution bandwidth to 4 MHz.
- Make sure system noise is capable of being measured.
- Correct for noise-near-noise.
- If a preamplifier is used, correct for amplifier noise figure.
- Insert a tunable bandpass filter if overload is suspected, and correct for the filter. losses. Correct for IF noise equivalent power bandwidth, NEPBW.

Selected Bibliography

- Bullinger, Rex. "Measurement of Differential Gain, Differential Phase, and Chrominance to Luminance Delay in Cable TV", Hewlett-Packard Company, Paper delivered at NCTA National Show, May 1994.
- *NCTA Recommended Practices For Measurements on Cable Television Systems*, 2nd ed. rev., October 1993.

Appendix E - Color-Video Measurements

Overview

FCC Regulations Part 73 set the compliance rules for U.S. broadcasters for color quality of their transmissions. In 1995 the rules governing cable television, FCC Regulations Part 76, include similar measurement standards that must be met by July 1 of that year. The three color measurements are:

- Chrominance-to-luminance delay inequality (CLDI).
- Differential gain (DG).
- Differential phase (DP).

This section introduces you to these color-quality tests in order to help you select the most effective measurement solution.

The spectrum analyzer, as defined for the measurements in this book, does not have the necessary circuits or computational power to make these measurements directly. The required analyzer modifications are outlined in brief. The advantages of an analyzer so equipped are that it can make many proof-of-performance measurements as well as video-color measurements in-service, that is, without disruption to subscriber service. The disadvantage is that the accuracy for the video tests, although sufficient for compliance, is not up to the broadcast laboratory test standards.

TIP

The spectrum analyzer alone cannot make CLDI, DG, or DP measurements, but when modified for the job, it can provide sufficient accuracy to make in-service compliance results.

Regulations

As mentioned, the FCC requires compliance by June of 1995. For chrominance-luminance delay inequality, the specification to be met is ±170 ns, where ns is nanoseconds, or 10^{-9} seconds. The differential gain specification is <20%, and differential phase is <10°. These units and levels should make sense after you read the rest of this appendix.

Differential Gain

TV Color Effect

If you watch a football game where the grass in the sunny side of the field looks washed out or more pale than the grass in the shade, chances are that the differential gain is out of specification. Differential gain is a measure of the change in the color saturation as the luminance, or brightness, changes.

Using a Test Signal

Measuring the change of gain in the video process as a function of luminance requires a test signal that offers several levels of luminance containing the color signal. The stair step or risers in one of several standard VITS provides such a signal sequence. This is illustrated in Figure 157.

The distance from zero voltage to the average of each packet is the luminance. The peak-to-peak height of each burst of chroma, at 3.58 MHz, is the magnitude of the signal. The heights of the packets are the same, representing a uniform gain for different levels of luminance. If video processing changes these burst heights, differential gain distortion results. The mathematical definition of differential gain is shown in the Figure 157.

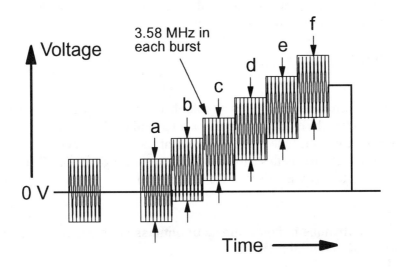

Figure 157. Packets of color burst signals are available in VITS for the measurement of differential gain.

Since the 3.58 MHz chroma packets are delivered at equal heights, noted as (a) through (e), representing uniform chroma magnitude. The differential gain is defined as the maximum gain of these minus the minimum gain, divided by the maximum, as shown in the following formula:

$$DG = [(max - min)/(max + 100)] \times 100$$

The packet magnitudes are normalized to the reference burst at blanking level and expressed as a percentage.

TIP   **Changes in color intensity, or saturation, as brightness changes is due to poor differential gain.**

The FCC requires that each channel in the system be below 20% differential gain. The stair step shown in Figure 157 is what you would see on a waveform monitor looking at a demodulated signal with either a full-field test signal for an off-line channel, or on a VITS inserted on an in-service broadcast signal. More will be said about measurement procedures in the last section of this appendix.

Differential Phase

TV Color Effect

The phase of a video signal determines its color hue, that is, its place in the color spectrum. Go back to the example of grass on a football field: Differential phase is out of specification if the grass is tinted yellow in sunlight and blue in shade. The luminosity is the brightness of the grass, in this case, low for shade and high for sun. The phase of the signal is specified to stay within 10° of the reference burst .

Changes in color hue as brightness changes is due to poor differential phase.

Using a Test Signal to Understand Effect

Phase is the position of a signal in time relative to another. The same stair-step test signal is used to measure phase changes as used to measure gain differences. The phase of each burst is referenced to the first burst measured. Whether the phase lags the reference or leads the reference is important because the differential phase is defined as the widest overall variance observed. As shown in Figure 158, each packet is compared in phase to the first packet on the left, the reference.

The formula for differential phase is simply the maximum phase shift minus the minimum phase shift in degrees. The specification limit is 10° of shift.

Measuring with a Spectrum Analyzer

The spectrum analyzer does not measure phase by itself. The good news is that it may have sufficient computing power to determine phase on the video signal using the fast Fourier transform function. Fast Fourier transform, or FFT, has the ability to extract the amplitude, phase, and harmonic frequency content of a time-varying signal.

To use the FFT, the analyzer requires extra circuits to trigger on the proper TV line, collect the video signal, and process the data for each packet. More on the specific requirements later.

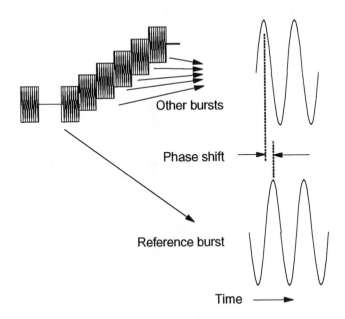

Figure 158. Phase of each burst compared to the reference burst to measure differential phase. Phases are greatly exaggerated to demonstrate the measurement definition.

Chrominance-to-Luminance Delay Inequality

TV Color Effect

This measurement is also known as CLDI, or chroma-lumina delay inequality, a difficult phrase to master. It is a measure of difference in the time it takes the chrominance and luminance parts of a signal to pass through the system.

The effect on a television receiver is poor horizontal registration of black and white, or luminance, and color information. Since the luminance and color information is scanned onto the picture tube horizontally, the timing differences between the arrival of the of the brightness and color signals results in edges that appear as colored shadows or ghosts. Each side of the edge will be a different color. An extreme case is illustrated in Figure 159. A dark red bar on a white background has the left edge gray, and the right edge bright red. The luminance information is arriving at the TV sooner than the color information, so each horizontal line displays three colors instead of the intended dark red.

The FCC requires that a delay or advance of no more than 170 ns, that is, 170×10^{-9} seconds delay between the luminance and chrominance signals, no matter which one leads the other. This is often written ±170 ns.

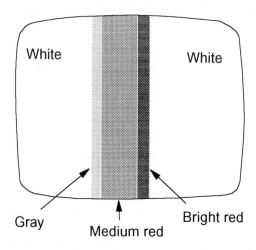

White   White

Gray   Bright red
        Medium red

Figure 159. Television representation of a dark red vertical bar on a white background is distorted by CLDI.

 TIP   **CLDI produces poor registration of the color and brightness on the TV picture.**

The 12.5T Pulse

It takes a special test signal to determine CLDI accurately. This signal, the so-called 12.5T pulse, is delivered along with other test signals in VITS. It is shaped specifically to produce two pulses in the video baseband, a luminance burst at the carrier frequency, and a color burst at the color subcarrier frequency. This is illustrated in Figure 160.

The luminance signal is transmitted as a carrier-frequency pulse with a carefully shaped envelope. The chrominance pulse, a burst of 3.58 MHz signal, is shaped similarly to the luminance pulse. The addition of these two pulses produces a single pulse whose voltage does not swing through zero. This is shown in Figure 160 as the straight base line in (a). In the frequency domain this combination pulse separates into its luminance and chrominance pulses as shown in (b). The chrominance pulse is 3.58 MHz above the luminance, just as is the color burst in the spectrum analyzer display of the color channel.

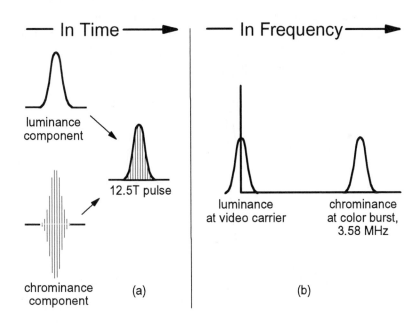

Figure 160. The 12.5T pulse launching a chrominance and luminance pulse. The pulse is shown in the time domain, (a), and frequency domain, (b).

TIP   **The 12.5T pulse is used to make CLDI measurements.**

Since the 12.5T pulse launches these two pulses at the same time, the luminance and chrominance should arrive at the TV receiver at the same time even after passing through modulators, processors, and converters. A delay of either lumina or chroma of more than 170 ns is considered poor color performance.

The delay is measured in the time domain by the shape of the baseline of the pulse voltage base as illustrated in Figure 161. In (a) no phase delay has occurred. In (b) the chrominance pulse is delayed, causing the summation of the two pulse amplitudes to cause the base line to rise at the leading edge and drop below zero on the lagging edge. In (b) the waveforms have been exaggerated to show the reason for this change in the base line. With the chrominance lagging behind the luminance, the amplitude sum of the points at the left side rise above the base-line since the lumina is always higher than the chroma response. In (c), where lumina lags behind chroma, just the opposite occurs. The energy of the chroma signal causes a dip in the base-line voltage initially.

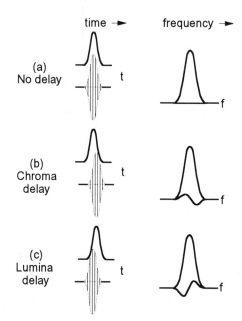

Figure 161. The shape of the 12.5T pulse after transmission indicates the type of phase delay.

Since the changes in base-line shape are dependent upon the gain of luminance and chrominance as well as their phase relationships, the base-line variations can be more complex that this. Detailed nomographs interpret the base-line amplitude and position wiggles into delay and gain quantities. The bibliography at the end of this appendix points to some these references.

The 12.5T pulse is broadcast with a built-in chroma advance. This advance is a pre-correction for the delay caused by the notch diplexer which is used to add the aural signal to the visual carrier at the transmitter's high power output. The pre-correction is 170 ns, the same amount as the compliance specification, so it must be accounted for in measuring CLDI.

Measurement Techniques

Spectrum Analyzer

The proof-of-performance tests in this book can be made with a spectrum analyzer purchased "off the shelf," with little or no modification as long as the analyzer has the minimum required measurement accuracy. Color tests, however, require additional functions not available in the standard spectrum analyzer. These are TV line triggered sweep, fast Fourier transform, and separate TV receiver. The TV line triggered sweep enables the analyzer to capture the appropriate portion of a broadcast vertical interval test signal, or VITS, for processing either in the time or frequency domains. The TV receiver provides an accurate trigger for extracting that data. Fast Fourier transform is a digital signal process on a portion of the VITS. This gives the analyzer the ability to measure changes in phase.

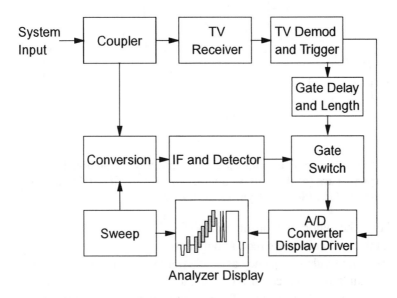

Figure 162. Simple block diagram of spectrum analyzer with added TV receiver.

TIP

A spectrum analyzer does not measure phase. To make these video measurements, the analyzer needs gated sweep, fast Fourier transform, and a separate TV receiver.

Test Signals Required

The second requirement is the need for a test signal transmitted along with the video under test. Either one of two specific VITS, the FCC Composite (also known as the Composite Radiated Signal) or the NTC-7 Composite test signal is used.

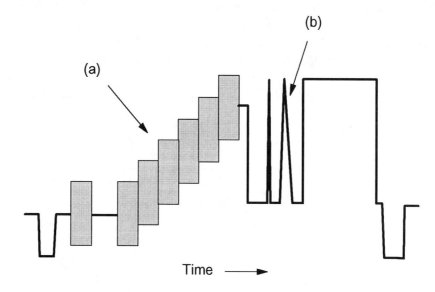

Figure 163. A typical composite video interval test signal. The responses are (a) the stair-step risers for differential gain and phase and (b) the 12.5 T pulses for CLDI.

Figure 163 shows the major components of a VITS. Several test responses are placed on the VITS, but the video color tests require only two of these. The stair-step signal (a) provides a variation of video frequency as a function of luminance, that is, amplitude. It also provides a standard reference for phase at each step, or riser, by which to measure phase change over levels of luminance. The pulse (b) cleverly provides a signal that determines the delay between the video and color signals, or chrominance-luminance delay inequalty.

If these signals are not provided by the broadcast source or injected at the cable system's head end, the analyzer cannot make color measurements. For example, the broadcaster may provide VIT signal in the feed, but commercial insertion equipment or data services may strip VITS and insert data signals or blank lines, requiring you to reinsert VITS. Broadcasting standards allow VITS to be stripped by processing amplifiers which reestablish sync and blanking. Often off-air UHF and satellite-delivered channels do not have VITS.

Waveform Monitor

The waveform monitor makes many more tests on VITS than the three compliance color tests discussed in this appendix. To make these tests on a waveform monitor, the RF TV signal is down converted to IF, demodulated to video baseband, and fed to the waveform monitor for quality analysis. Figure 164 shows a simple block diagram for this measurement.

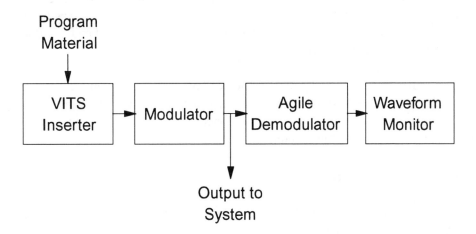

Figure 164. Waveform monitor requires a converter and high-quality demodulator to measure video color parameters.

The combination of a precision down converter/demodulator and waveform monitor is more accurate than the spectrum analyzer, but the trade off is in cost, and, for a customized spectrum analyzer solution, the benefit is the convenience of testing. You will pay a higher price for the additional accuracy of the waveform monitor, down converter, and demodulator.

Overall Accuracy Considerations

In compliance testing a general rule is to have your test equipment accurate to within 1/10 to 1/4 of the specifications under test. For the color-video measurements these are:

- ◆ 2% to 5% for differential gain.
- ◆ 1° to 2.5° for differential phase.
- ◆ ±17 ns to ±43 ns for CLDI.

Refer to the specifications of the test equipment to make sure compliance level accuracy is met.

The FCC Composite or NTC-7 Composite VITS provide the necessary test signals for color-video measurements with either a customized spectrum analyzer or waveform monitor. Keep in mind that these VITS are subject to transmission distortions from signal multipath at the head end antenna. So be sure to test the video at head end input before testing post-processing distribution points.

The waveform analyzer is usually much more accurate than the customized spectrum analyzer, but the trade off with the waveform analyzer is that the measurement usually requires operator interpretation which can add uncertainty to the measurement and be more difficult to record for compliance auditing.

Selected Bibliography

- ◆ Benson, K. Blair, and Whitaker, Jerry. *Television Engineering Handbook*. Rev. ed., McGraw-Hill, Inc., 1992.
- ◆ Bullinger, Rex. "Measurement of Differential Gain, Differential Phase, and Chrominance to Luminance Delay in Cable TV", Hewlett-Packard Company, Paper delivered at NCTA National Show, May 1994.
- ◆ Craig, Margaret. *Television Measurements - NTSC Systems*. Tektronix Television Division, Beaverton, OR, 1989.
- ◆ Edgington, Francis M. "Preparing for In-service Video Measurements", *Communications Engineering & Design*, June 1994.

- Inglis, Andrew F. *Video Engineering*. McGraw-Hill, Inc., New York, 1993.
- *NCTA Recommended Practices For Measurements on Cable Television Systems*, 2nd ed. rev., October 1993.
- Vartanian, John. *Tech Rereg: Differential Gain, Differential Phase, Chrominance-to-Luminance Delay Inequality*. Communications Technology Publications Corp., October 1993.
- Webb, Jack. *Video Testing Step by Step - Part 1*. Communications Technology Publications Corp., June 1994.

Appendix F - Glossary of CATV and Spectrum Analyzer Terms

active trace In spectrum analyzer operation, the trace (commonly A, B, or C) that is being swept (updated) with incoming signal information.

active tap A cable television feeder device consisting of a directional coupler and a hybrid splitter (e.g., a conventional subscriber tap), in addition to an amplifier circuit.

adapter Mechanism for attaching parts, especially those parts having different physical dimensions or electrical connectors.

adjacent channel (1) Any two television channels spaced 6 MHz apart. (2) The channel (frequency band) immediately above or below the channel of interest.

ambient temperature The temperature surrounding apparatus and equipment. Synonymous with room temperature.

amplifier Device used to increase the operating level of an input signal. Used in a cable system's distribution plant to compensate for the effects of attenuation caused by coaxial cable and passive device losses.

amplitude The size or magnitude of a voltage or current waveform; the strength of a signal.

amplitude accuracy In spectrum analyzer operation, the general uncertainty of a spectrum analyzer amplitude measurement, whether relative or absolute.

amplitude The size or magnitude of a voltage or current waveform; the strength of a signal.

amplitude modulation (AM) The form of modulation in which the amplitude of the signal is varied in accordance with the instantaneous value of the modulating signal.

amplitude modulated link (AML) A form of microwave communications using amplitude modulation for the transmission of television and related signals.

analog Pertaining to signals in the form of continuously variable physical quantities.

antenna Any structure or device used to collect or radiate electromagnetic waves.

antenna array A radiating or receiving system composed of several spaced radiators or elements.

attenuation The difference between transmitted and received power due to loss through equipment, lines, or other transmission devices; usually expressed in decibels. In spectrum analyzer operation, a general term used to denote a decrease of signal magnitude in transmission from one point to another. Attenuation may be expressed as a scalar ratio of the input to the output magnitude in decibels.

attenuator A device for reducing the amplitude of a signal.

audio Relating to sound or its reproduction; used in the transmission or reception of sound.

audio frequency A frequency lying within the audible spectrum (the band of frequencies extending from about 20 Hz to 20 kHz).

aural carrier The carrier that has the audio portion of a television channel. A television channel usually contains both a visual and an aural carrier. An aural carrier is sometimes referred to as a sound carrier.

aural center frequency (1)The average frequency of an emitted signal when modulated by an aural (audio) signal. (2) The frequency of the emitted wave without modulation. Usually refers to frequency modulation methods.

automatic tilt Automatic correction of changes in tilt, or the relative level of signals of different frequencies.

B

bandpass filter A device that allows signal passage to frequencies within its design range and that effectively bars passage to all signals outside that frequency range.

bandwidth (1) A measure of the information carrying capacity of a communication channel. The bandwidth corresponds to the difference between the lowest and highest frequency signal that can be carried by the channel. (2)

The range of usable frequencies that can be carried by a cable television system.

bandwidth selectivity A measure of the spectrum analyzer's ability to resolve signals unequal in amplitude. It is the ratio of the 60 dB bandwidth to the 3 dB bandwidth for a given resolution filter (IF). Bandwidth selectivity tells us how steep the filter skirts are. Bandwidth selectivity is sometimes called the shape factor.

baseband The band of frequencies occupied by the signal in a carrier wire or radio transmission system before it modulates the carrier frequency to form the transmitted line or radio signal.

baseband channel Connotes that modulation is used in the structure of the channel, as in a carrier system. The usual consequence is phase or frequency offset. The simplest example is a pair of wires that transmits direct current and has no impairments such as phase offset or frequency offset that would destroy waveform.

black level That level of picture signal corresponding to the maximum limit of black peaks.

bridging amplifier An amplifier connected directly into the main trunk of the cable television system. It serves as a sophisticated tap, providing isolation from the main trunk, and has multiple high-level outputs that provide signals to the feeder portion of the distribution network. Synonymous with bridger and distribution amplifier.

broadband Any system able to deliver multiple channels and/or services to its users or subscribers. Generally refers to cable television systems. Synonymous with wideband.

C

cable television A broadband communications technology in which multiple television channels as well as audio and data signals are transmitted either one way or bidirectionally through a distribution system to single or multiple specified locations. The term also encompasses the associated and evolving programming and information resources that have been and are being developed at the local, regional, and national levels.

cable television relay station (CARS) A fixed or mobile microwave communications station used for the transmission of television and related audio signals, FM broadcast stations, cablecasting, data or other information, or test signals for reception at one or more fixed receiving points from which the signals are then distributed to the public by cable.

cable television system A broadband communications system capable of delivering multiple channels of entertainment programming and non-entertainment information from a set of centralized antennas, generally by coaxial cable, to a community. Many cable television designs integrate microwave and satellite links into their overall design, and some now include optical fibers as well. Often referred to as cable television, which usually stands for community antenna television system.

carrier An electromagnetic wave of which some characteristic is varied in order to convey information.

carrier-to-noise ratio The ratio of amplitude of the carrier to the noise power relative to a 4 MHz bandwidth in the portion of the spectrum occupied by the carrier. Also referred to as the C/N ratio, or C/N.

CATV Abbreviation for community antenna television or cable television system. A cable television system is a broadband communications system that provides multiple channels from centralized antennas.

center frequency (1) The average frequency of the emitted wave when modulated by a sinusoidal wave. (2) The frequency of the emitted wave without modulation.

certificate of compliance Authorization issued by the FCC for the operation of a cable television system in a community or for the carriage of additional television signals by an operating cable television system.

channel A signal path of specified bandwidth for conveying information.

channel capacity In a cable television system, the number of channels that can be simultaneously carried on the system. Generally defined in terms of the number of 6 MHz (television bandwidth) channels for NTSC.

channel frequency response (1) The relationship within a cable television channel between amplitude and frequency of a constant amplitude input signal as measured at a subscriber terminal. (2) The measure of amplitude frequency distortion within a specified bandwidth.

coaxial cable A type of cable used for broadband data and cable systems. Composed of a center conductor, insulating dielectric, conductive shield, and optional protective covering, this type of cable has excellent broadband frequency characteristics, noise immunity, and physical durability. Synonymous with coax.

co-channel interference Interference on a channel caused by another signal operating on the same channel.

Community Antenna Television System See **CATV**.

composite The effect of several distortion signals present within a very narrow bandwidth. See discrete.

composite second-order beat (CSO) (1) A clustering of second order beats 1.25 MHz above the visual carriers in cable systems. (2) A ratio, expressed in decibels, of the peak level of the visual carrier to the peak of the average level of the cluster of second-order distortion products located 1.25 MHz above the visual carrier.

composite triple beat (CTB) (1) A clustering of third-order distortion products around the visual carriers in cable systems. (2) A ratio, expressed in decibels, of the peak level of the visual carrier to the peak of the average level of the cluster of third-order distortion products centered around the visual carrier.

compression (1) A less-than-proportional change in output for a change in input. (2) The reduction in amplitude of one portion of a waveform relative to another portion.

continuous sweep mode The spectrum analyzer condition where traces are automatically updated each time trigger conditions are met.

converter Also known as processor. Device for changing the frequency of a television signal. A cable head end converter changes signals from frequencies at which they are broadcast to

clear channels that are available on the cable distribution system. A set-top converter is added in front of a subscriber's television receiver to change the frequency of the midband, superband, or hyperband signals to a suitable channel or channels (typically a low-VHF channel) which the television receiver is able to tune.

cross modulation A form of television signal distortion where modulation from one or more television channels is imposed on another channel or channels.

CSO See **composite second-order beat**.

CTB See **composite triple beat**.

D

dB See **decibel**.

dBc Decibel carrier. A ratio expressed in decibels that refers to the gain or loss relative to a reference carrier level.

dBm See **decibel milliwatt**.

dBmV See **decibel millivolt**.

decibel (dB) A unit that expresses the ratio of two power levels on a logarithmic scale.

decibel millivolt (dBmV) A unit of measurement referenced to one millivolt across a specified impedance (75 ohms in cable television).

decibel milliwatt (dBm) A unit of measurement referenced to one milliwatt across a specified impedance.

delta marker A spectrum analyzer mode in which a fixed reference marker is established, then a second active marker becomes available so it can be placed anywhere along the trace. A readout indicates the relative frequency separation and amplitude difference between the reference and active markers.

demodulate To retrieve an information-carrying signal from a modulated carrier.

demodulator A device that removes the modulation from a carrier signal.

discrete Distortion and interference signals generated from a combination of known tones. Also see **composite.**

display dynamic range The maximum dynamic range over which both the larger and smaller signals can be viewed simultaneously on the display. For spectrum analyzers with a maximum logarithmic display of 10 dB/division, the actual dynamic range may be greater than the display dynamic range. See **dynamic range.**

display fidelity In spectrum analyzer operation, the measurement uncertainty of relative differences in amplitude. With digital displays, markers are used to measure the signal. As a result, measurement differences are stored in memory, and the ambiguity of the display is eliminated from the measurement.

distortion An undesired change in waveform of a signal in the course of its passage through a transmission system.

distribution amplifier See **bridging amplifier.**

distribution system The part of a cable television system consisting of trunk and feeder cables that are used to carry signals from the system head end to subscriber terminals. Often applied, more narrowly, to the part of a cable television system starting at the bridger, amplifiers. Synonymous with trunk and feeder system.

downconverter A type of radiofrequency converter characterized by the frequency of the output signal being lower than the frequency of the input signal. Synonymous with input converter.

drift A change in the output of a circuit that occurs slowly.

dynamic range In general, the ratio (in decibels) of the weakest or faintest signals to the strongest or loudest signals reproduced without significant noise or distortion. In a spectrum analyzer, the power ratio (dB) between the smallest and largest signals simultaneously present at the input that can be measured with some degree of accuracy. Dynamic range generally refers to measurement of distortion or intermodulation products.

E

electromagnetic interference Any electromagnetic energy, natural or man-made, which may adversely affect performance of the system.

electromagnetic spectrum The frequency range of electromagnetic radiation that includes radio waves, light and X-rays. At the low frequency end are subaudible frequencies (i.e., 10 Hz) and at the other end, extremely high frequencies (e.g., X-rays, cosmic rays).

F

FFT The abbreviation for fast Fourier transform. It is a mathematical operation performed on a time-domain signal to yield the individual spectral components that constitute the signal.

fiber optics The technology of guiding and projecting light for use as a communications medium. Hair-thin glass fibers that allow light beams to be bent and reflected with low levels of loss and interferences are known as "glass optical wave guides" or simply "optical fibers."

field One-half of a complete picture (or frame) interval, containing all of the odd or even scanning lines of the picture.

field frequency The rate at which a complete field is scanned, nominally 60 times per second for NTSC monochrome video signals, and 59.94 times per second for NTSC color video signals .

field strength The intensity of an electromagnetic field at a given point, usually referred to in microvolts per meter.

field strength meter (FSM) A frequency selective heterodyne receiver capable of tuning to the frequency band of interest; in cable television, 5 to 750 MHz and above with indicating meter showing the magnitude input of voltage and a dial indicating the approximate frequency. Synonymous with signal level meter.

FM modulator In cable television, a device similar to an FM transmitter that is used to cablecast signals in the FM band on a cable system.

Fourier series A mathematical analysis permitting any complex waveform to be resolved into a fundamental plus a finite number of terms involving its harmonics.

frame One complete picture consisting of two fields of interlaced scanning lines.

frequency accuracy In spectrum analyzer operation, the uncertainty with which the frequency of a signal or spectral component is indicated, either in an absolute sense or

relative to some other signal or spectral component. Absolute and relative frequency accuracies are specified independently.

frequency band splitter/mixer A device similar to other splitters except that it provides branching on a frequency division basis.

frequency modulation (FM) A form of modulation in which the frequency of the carrier is varied in accordance with the instantaneous value of the modulating signal.

frequency range The range of frequencies over which the spectrum analyzer performance is specified. The maximum frequency range of many microwave spectrum analyzers can be extended with the application of external mixers.

frequency reference error In spectrum analyzers, frequency reference error is the amount of specified change allowed of the analyzer's built-in standard oscillator, often a crystal-controlled oscillator or synthesizer. The values in this specification are part of the analyzer's frequency accuracy equations. Often the frequency reference error is given for a certain time period, called aging, such as the calibration cycle of one year, or a temperature stability, specifying how the total drift over the analyzer's operating temperature, such as 0°C to 55°C. Stability is the ability of a frequency component to remain unchanged in frequency or amplitude over time.

frequency resolution The ability of a spectrum analyzer to separate closely spaced spectral components and display them individually. Resolution of equal amplitude components is determined by resolution bandwidth. Resolution of unequal amplitude signals is determined by resolution bandwidth and bandwidth selectivity.

frequency response The peak-to-peak variation in the displayed signal amplitude over a specified center frequency range. Frequency response is typically specified in terms of ±dB relative to the value midway between the extremes. It also may be specified relative to the calibrator signal.

frequency span In spectrum analyzer operation, the magnitude of the displayed frequency component. Span is represented by the horizontal axis of the display. Generally, frequency span is given as the total span across the full display. Some spectrum analyzers represent frequency span (scan width) as a per-division value.

frequency stability Stability is the ability of a frequency component to remain unchanged in frequency or amplitude over short- and long-term periods of time. In spectrum analyzers, stability refers to the local oscillator's ability to remain fixed at a particular frequency over time. The sweep ramp that tunes the local oscillator influences where a signal appears on the display. Any long-term variation in local oscillator frequency (drift) with respect to the sweep ramp causes a signal to shift its horizontal position on the display slowly. Shorter-term local oscillator instability can appear as random FM or phase noise on an otherwise stable signal.

front-panel key In spectrum analyzer operation, keys, typically labeled, located on the front panel of an instrument. The key labels identify the function and the key activities. Numeric keys and step keys are two examples of front-panel keys.

full span A mode of operation in which the spectrum analyzer scans the entire frequency band of an spectrum analyzer.

function The action or purpose that a specific item is intended to perform or serve. The spectrum analyzer contains functions that can be executed via front-panel key selections, or through programming commands. The characteristics of these functions are determined by the firmware in the instrument. In some cases, a DLP (downloadable program) execution of a function allows you to execute the function from front-panel key selections.

G

gain compression The signal level at the input mixer of a spectrum analyzer where the displayed amplitude of the signal is a specific number of dB too low due just to mixer saturation. The signal level is generally specified for 1 or 0.5 dB compression and is usually between −3 and −10 dBm.

graticule line (reference level), and scale factor in volts per division. On most spectrum analyzers, the bottom graticule line represents 0 (zero) volts for scale factors of 10 dB/division or more. The bottom division, therefore, is not calibrated for those spectrum analyzers. Spectrum analyzers with microprocessors allow reference level and marker values to be indicated in dBm, dBmV, dBmV, volts, and occasionally in watts.

H

harmonic distortion (1) The generation of harmonics by the circuit or device with which the signal is processed. (2) Unwanted harmonic components of a signal .

harmonically related carriers Harmonically related carriers (HRC) is a tune configuration where each video carrier is a multiple of 6 MHz. This configuration masks composite triple-beat distortion by zero-beating the composite triple-beat distortion with the video carriers.

head end The control center of a cable television system, where incoming signals are amplified, converted, processed, and combined into a common cable, along with any origination cablecasting, for transmission to subscribers. System usually includes antennas, preamplifiers, frequency converters, demodulators, modulators, processors, and other related equipment.

herringbone An interference pattern in a television picture, appearing as either moving or stationary rows of parallel diagonal or sloping lines superimposed on the picture information.

hertz (Hz) A unit of frequency equivalent to one cycle per second.

heterodyne To mix two frequencies together in a nonlinear component in order to produce two other frequencies equal to the sum and difference of the first two. Synonymous with beat.

heterodyne processor An electronic device used in cable head ends that down-converts an incoming signal to an intermediate frequency for filtering, signal level control, and other processing, and then reconverts that signal to a desired output frequency.

HRC See **harmonically related carriers.**

hum modulation Undesired modulation of the television visual carrier by power-line frequencies or their harmonics (e.g., 60 or 120 Hz), or other low-frequency disturbances.

Hz See **hertz.**

I

impedance The combined effect of resistance, inductive reactance, and capacitive reactance on a signal at a particular frequency. In cable television, the nominal impedance of the cable and components is 75-ohms.

impedance matching A method used to match two or more components into a single characteristic impedance of one of the components, to minimize attenuation and anomalies.

incremental coherent carriers (ICC/IRC) A cable plan in which all channels except 5 and 6 correspond with the standard channel plan. The technique is used to reduce composite triple-beat distortions. Synonymous with incrementally related carriers.

incrementally related carriers Incrementally related carriers (IRC) is a tune configuration where all channels except channels 5 and 6 are standard channels (see **standard tune configuration** for a definition of standard channels).

ingress The unwanted leakage of interfering signals into a cable television system.

input attenuator An attenuator (also called an RF attenuator) between the input connector and the first mixer of a spectrum analyzer. The input attenuator is used to adjust the signal level incident to the first mixer, and to prevent gain compression due to high-level or broadband signals. It is also used to set the dynamic range by controlling the degree of internally generated distortion. For some spectrum analyzers, changing the input attenuator settings changes the vertical position of the signal on the display, which then changes the reference level accordingly.

input impedance The terminating impedance that the spectrum analyzer presents to the signal source. The nominal impedance for RF and microwave spectrum analyzers is usually 50-ohms. For some systems, such as cable TV, 75-ohms is standard. The degree of mismatch between the nominal and actual input impedance is called the VSWR (voltage standing wave ratio).

insertion loss Additional loss in a system when a device such as a directional coupler is inserted, equal to the difference in signal level between input and output of such a device.

insertion test signals See **vertical interval reference test signal**.

Institute of Electrical and Electronic Engineers (IEEE) An engineering society formed by the merger of the Institute of Radio Engineers and the American Institute of Electrical Engineers.

IRC See **incrementally related carriers**.

L

leakage Undesired emission of signals out of a cable television system, generally through cracks in the cable, corroded or loose connections, or loose device closures. Synonymous with signal leakage.

line extender Feeder line amplifiers used to boost signal and thereby extend the useful range of the feeder cable.

line frequency (1) The number of times per second that the scanning spot crosses a fixed vertical line in one direction. (2) Related to commercial power line frequency, i.e., 60 Hz.

(3) The horizontal scanning rate of a video signal. For NTSC video, 15.734 kHz.

local oscillator An oscillator, built into the design of the equipment, that generates a signal used in the heterodyne process to mix with incoming signals and produce an intermediate frequency .

log display In spectrum analyzer operation, the display mode in which vertical deflection is a logarithmic function of the input-signal voltage. Log display is also referred to as logarithmic mode. The display calibration is set by selecting the value of the top graticule.

low-frequency interference Interference effects that occur at low frequency, generally considered as any frequency below 15.734 kHz.

low noise amplifier (LNA) A low noise signal booster used to amplify the weak signals received on a satellite antenna.

low noise block converter A combination device used on satellite antennas that includes both a low noise amplifier (LNA) to boost the weak signals, and a block downconverter to convert the incoming satellite signals to a lower band of frequencies (e.g., 70-1450 MHz).

low pass filter (LPF) A filter which passes all frequencies below a specified frequency, and blocks those frequencies above the specified frequency.

M

main trunk The major cable link or "backbone" from the headend to a community or between communities.

marker In spectrum analyzer operation, a visual indicator placed anywhere along the displayed trace. A marker readout indicates the absolute value of the trace frequency and amplitude at the marked point. The amplitude value is displayed with the currently selected units.

matching transformer An impedance matching device which converts the 75-ohm impedance of the subscriber drop to the 300-ohm impedance of a television or FM receiver.

maximum input level In spectrum analyzer operation, the maximum signal power that may be safely applied to the input of a spectrum analyzer. Typically 1 W (–30 dBm) for Hewlett-Packard spectrum analyzers.

measurement range In spectrum analyzer operation, the ratio, expressed in dB, of the maximum signal level that can be measured (usually the maximum safe input level) to the lowest achievable average noise level. This ratio is almost always much greater than can be realized in a single measurement. Refer also to dynamic range.

megahertz (MHz) One million cycles per second.

menu The spectrum analyzer functions that appear on the display and are selected by pressing front-panel keys. These selections may evoke a series of other related functions that establish groups called menus.

microsecond One millionth of a second.

microwave A very short wavelength electromagnetic wave, generally above 1000 MHz.

mismatch (1)The condition resulting from connecting two circuits or connecting a line to a circuit in which the two impedances are different. (2) Impedance discontinuity.

modulate To vary the amplitude, frequency, or phase of a carrier or signal in accordance with the instantaneous amplitude and/or frequency changes of the modulating intelligence.

modulation The process whereby original information can be translated and transferred from one medium to another. Information originally carried as a variation in a particular property (such as amplitude) of one process is transferred and carried as a corresponding variation in some possible different property (such as duration) of the new process.

N

narrow band A relative term referring to a system that carries a narrow- frequency range (sometimes used to refer to frequency bandwidths below 1 MHz). In a telephone/television context, telephone would be considered narrow band (3 kHz) and television would be considered broadband (6 MHz).

National Cable Television Association (NCTA) Washington, D.C.-based trade association for the cable television industry; members are cable television system operators; associate members include cable hardware and program suppliers and distributors, law and brokerage firms, and financial institutions. NCTA represents the cable television industry before state and federal policy makers and legislators. Name was changed in 1969 from National Community Television Association.

NCTA See **National Cable Television Association**.

noise Random burst of electrical energy or interference which may produce a "salt-and-pepper" pattern over a television picture. Heavy noise is sometimes called "snow."

noise factor Ratio of input signal-to-noise ratio to output signal-to-noise ratio.

noise figure The amount of noise added by signal-handling equipment (e.g., an amplifier) to the noise existing at its input, usually expressed in decibels.

noise temperature The temperature that corresponds to a given noise level from all sources, including thermal noise, source noise, and induced noise.

NTSC video signal A 525-line color-video signal whose frequency spectrum extends from 30 Hz to 4.2 MHz. NTSC video consists of 525 interlaced lines, with a horizontal scanning rate of 15,734 Hz, and a vertical (field) rate of 59.94 Hz. A color subcarrier at 3.579545 MHz contains color hue (phase) and saturation (amplitude) information .

O

off-the-air tune configuration The tune configuration for signals that are broadcast over the air and received with an antenna.

oscilloscope An oscillograph test apparatus primarily intended to visually represent test or troubleshooting voltages with respect to time. Synonymous with scope.

P

pass band The range of frequencies passed by a filter, amplifier, or electrical circuit.

passive device A device basically static in operation; that is, it is not capable of amplification or oscillation, and requires no power for its intended function. Examples include splitters, directional couplers, taps, and attenuators.

peak power The power over a radiofrequency cycle corresponding in amplitude to synchronizing peaks. Refers to television broadcast transmitters.

percentage modulation (amplitude) The ratio of half the difference between the maximum and minimum amplitudes of an amplitude-modulated wave to the average amplitude expressed in percentage.

percentage modulation (FM) As applied to frequency modulation (1) the ratio of the actual frequency swing defined as 100% modulation, expressed in percentage. (2) The ratio of half the difference between the maximum and minimum frequencies of the average frequency of an FM signal.

performance standards Certain minimum technical requirements, established by the appropriate regulatory body, which must be met by a cable system operator.

pilot carrier Signals on cable television systems used to operate attenuation (gain) and frequency response (slope) compensating circuitry in amplifiers.

pilot subcarrier A subcarrier serving as a control signal for use in the reception of stereophonic broadcasts.

preamplifier A low-noise electronic device (usually installed near an antenna) designed to strengthen or boost a weak off-air signal to a level where it will overcome antenna download loss and be sufficient to drive succeeding processors or amplifiers. In spectrum analyzers, an external, low-noise-figure amplifier that improves system spectrum analyzer (preamplifier/spectrum analyzer) sensitivity over that of the spectrum analyzer itself.

pulse A variation in the value of a quantity, short in relation to the time schedule of interest, with the final value being the same as the initial value.

R

radiofrequency (RF) An electromagnetic signal above the audio and below the infrared frequencies.

random noise Thermal noise generated from electron motion within resistive elements of electronic equipment.

reference level In spectrum analyzer operation, the calibrated vertical position on the display used as a reference for amplitude measurement in which the amplitude of one signal is compared with the amplitude of another regardless of the absolute amplitude of either.

resolution See **frequency resolution**.

resolution A measure of picture-resolving capabilities of a television system determined primarily by bandwidth, scan rates, and aspect ratio. Relates to fineness of details perceived.

resolution (horizontal) The amount of resolvable detail in the horizontal direction in a picture. It is usually expressed as the number of distinct vertical lines, alternately black and white, that can be seen in three-quarters of the width of the picture. This information usually is derived by observation of the vertical edge of a test pattern. A picture that is sharp and clear and shows small details has good, or high, resolution. If the picture is soft and blurred and small details are indistinct, it has poor, or low, resolution. Horizontal resolution depends upon

the high-frequency amplitude and phase response of the pickup equipment, the transmission medium, and the picture monitor, as well as on the size of the scanning spots.

resolution (vertical) The amount of resolvable detail in the vertical direction in a picture. It is usually expressed as the number of distinct horizontal lines, alternately black and white, that can be seen in a test pattern. Vertical resolution is primarily fixed by the number of horizontal scanning lines per frame. Beyond this, vertical resolution depends on the size and shape of the scanning spots of the pickup equipment and picture monitor and does not depend upon the high-frequency response or bandwidth of the transmission medium or picture monitor.

resolution bandwidth This term is used to identify the width of the resolution bandwidth filter of a spectrum analyzer. The 3 dB resolution bandwidth is specified; for others, it is the 6 dB resolution bandwidth. Resolution bandwidth relates to the ability of a spectrum analyzer to display adjacent responses discretely.

response time The time interval between the instant a signal or stimulus is applied to or removed from a device or circuit, and the instant the circuit or device responds or acts.

S

scrambled To alter an electronic signal so that a decoding device is necessary to receive the signal.

second harmonic In a complex wave, a signal component whose frequency is twice the fundamental, or original, frequency.

second-order beat Even-order distortion product created by two signals mixing or beating against each other.

shape factor See **bandwidth selectivity**.

sidebands Additional frequencies generated by the modulation process, which are related to the modulating signal and contain the modulating intelligence.

signal generator An electronic instrument that produces audio- or radiofrequency signals for test, measurement, or alignment purposes.

signal leakage See **leakage**.

signal level Amplitude of signal voltage measured across 75-ohms, usually expressed in decibel millivolts.

signal level meter (SLM) See **field strength meter**.

signal-to-noise ratio The ratio, expressed in decibels, of the peak voltage of the signal of interest to the root-mean-square voltage of the noise in that signal.

single-sweep mode The spectrum analyzer sweeps once when trigger conditions are met. Each sweep is initiated by pressing an appropriate front-panel key, or by sending a programming command.

softkey In spectrum analyzer operation, key labels displayed on a screen or monitor that are activated by mechanical keys surrounding the display or located on a keyboard. Softkey selections usually evoke menus that are written into the program software. Front-panel key selections determine which menu (set of softkeys) appears on the display.

span In spectrum analyzer operation, span equals the stop frequency minus the start frequency. The span setting determines the horizontal-axis scale of the spectrum analyzer display.

span accuracy In spectrum analyzer operation, the uncertainty of the indicated frequency separation of any two signals on the display.

spectrum analyzer A scanning receiver with a display that shows a plot of frequency versus amplitude of the signals being measured. Modern spectrum analyzers are often microprocessor controlled and feature powerful signal-measurement capabilities.

splitter Usually a hybrid device, consisting of a radiofrequency transformer, capacitors, and resistors, that divides the signal power from an input cable equally between two or more output cables.

spurious signals Any undesired signals such as images, harmonics, and beats.

standard tune configuration The tune configuration in which the channels are at the frequencies that the Electronic Industries Association (EIA) and FCC define to be the standard channel frequencies.

STD See **standard tune configuration**.

stop/start frequency In spectrum analyzer operation, terms used in association with the stop and start points of the frequency measurement range. Together they determine the span of the measurement range.

subcarrier A carrier used to modulate information upon another carrier, for example, the difference channel subcarrier in an FM stereo transmission.

suckout (1) The result of the coaxial cable's center conductor, and sometimes the entire cable, being pulled out of a connector because of contraction of the cable. (2) A sharp reduction of the amplitude of a relatively narrow group of frequencies within the cable system's overall frequency response.

sweep generator An electronic instrument whose output signal varies in frequency between two preset or adjustable limits, at a rate that is also adjustable. This "swept" signal is used to perform frequency-response measurements when used in conjunction with appropriate peripheral accessories.

sweep time In spectrum analyzer operation, the time it takes the local oscillator to tune across the selected span. Sweep time directly affects how long it takes to complete a measurement.

sync generator An electronic device that supplies common synchronizing signals to a system of several video cameras, switchers, and other video production equipment, ensuring that all will be "locked" to a master timing reference.

sync level The level of the tips of the synchronizing pulses, usually 40 IRE units from blanking to sync tip.

synchronization The maintenance of, one operation in frequency and/or in phase with another.

synchronization pulse A transmitted pulse that is used to synchronize the electron beam of a picture monitor with the scanning device of the transmission source.

synchronous For transmission, operation of the sending and receiving instruments continuously at the same frequency .

system impedance The resistance and reactance opposing the current flow in the system. For cable television the impedance is 75 ohms. See also **impedance**.

system noise That combination of undesired and fluctuating disturbances within a cable television channel that degrades the transmission of the desired signal.

T

television channel The range or band of the radiofrequency spectrum assigned to a television station. In Canada and the United States, the standard bandwidth is 6 MHz.

terminal isolation The attenuation, at any subscriber terminal, between that terminal and any other subscriber terminal in the cable television system.

third harmonic In a complex wave, a signal component whose frequency is three times the fundamental, or original, frequency.

third-order beat See **triple beat**.

tilt compensation The action of adjusting, manually or automatically, amplifier frequency/gain response to compensate for different cable length frequency/attenuation characteristics.

trap (1) A passive device used to block a channel or channels from being received by a cable television subscriber (negative trap), or used to remove an interfering carrier from a channel that a subscriber wants to receive (positive trap). (2) An unprogrammed, conditional jump to a specified address that is automatically activated by hardware, a recording being made of the location from which the jump occurred.

triple beat Odd order distortion products created by three signals, mixing or beating against each other, whose frequencies fall at the algebraic sums of the original frequencies. Synonymous with third-order beat.

trunk The main distribution lines leading from the head end of the cable television system to the various areas where feeder lines are attached to distribute signals to subscribers.

trunk amplifier An amplifier inserted into a trunk line. A weak input signal is amplified before being retransmitted to an output line, usually carrying a number of video voice or data channels simultaneously. Amplifiers increase the range of a system. Usually, trunk amplifiers must be inserted approximately every 1,500 to 2,000 feet.

tune configuration Refers to the correlation between the channel numbers and the frequency to which a channel is assigned. For example, the frequency at which channel 1 is broadcast differs between the HRC and IRC tune configurations.

U

units Dimensions on the measured quantities. Units usually refer to amplitude quantities because they can be changed. In spectrum analyzers with microprocessors, available units are dBm (dB relative to 1 mW dissipated in the nominal input impedance of the spectrum analyzer), dBmV (dB relative to 1 mV), dBV (dB relative to 1 V), volts, and, in some spectrum analyzers, watts.

unscrambled A signal that has not been scrambled. An unscrambled signal does not need a decoder to receive the signal correctly.

V

vertical interval test signal (VITS) A signal that may be included during the vertical blanking interval to permit in-service testing and adjustment of video transmission.

vestigial sideband Amplitude modulation in which the higher frequencies of the lower sideband are not transmitted. At lower baseband frequencies, the carrier envelope is the same as that for normal double-sideband AM.

video A term pertaining to the bandwidth and spectrum of the signal that results from television scanning and that is used to reproduce a picture. In spectrum analyzer operation, a term describing the output of a spectrum analyzer's envelope detector. The frequency range extends from 0 Hz to a frequency that is typically well beyond the widest resolution bandwidth available in the spectrum analyzer. However, the ultimate bandwidth of the video chain is determined by the setting of the video filter. Video is also a term describing the television signal composed of a visual and aural carriers.

video average The digital averaging of spectrum analyzer trace information. It is available only on spectrum analyzers with digital displays.

video bandwidth (1) The maximum rate at which dots of illumination are displayed on a screen. (2) The occupied bandwidth of a video signal. For NTSC, that bandwidth is 4.2 MHz. (3) In spectrum analyzer operation, the cut-off frequency (3 dB point) of an adjustable low-pass filter in the video circuit. When the video bandwidth is equal to or less than the resolution bandwidth, the video circuit cannot fully respond to the more rapid fluctuations of the output of the envelope detector. The result is a smoothing of the trace seen as a reduction in the peak-to-peak noise excursion. The degree of averaging or smoothing is a function of the ratio of the video bandwidth to the resolution bandwidth.

video filter In spectrum analyzer operation, a post-detection, low-pass filter that determines the bandwidth of the video amplifier. It is used to average or smooth a trace. Refer also to video bandwidth.

visual carrier The visual carrier is the portion of a television signal that contains the picture. A television signal contains both a visual and an aural carrier.

visual carrier frequency The frequency of the carrier that is modulated by the picture information, which is 1.25 MHz above the bottom end of a television channel.

visual signal level The peak voltage produced by the visual signal during the transmission of synchronizing pulses.

W

waveform monitor A special-purpose oscilloscope which presents a graphic illustration of the video and sync signals, amplitude, and other information used to monitor and adjust baseband video signals.

white level The level of a visual carrier that corresponds to the maximum level of the white area for a picture signal.

windshield wiper effect Onset of overload in multichannel cable television systems caused by cross modulation; the horizontal sync pulses of one or more television channels are superimposed on the desired channel carrier. The visual effect of the interference resembles a diagonal bar wiping through the picture.

Z

zero span In spectrum analyzer operation, the local oscillator remains fixed at a given frequency so that the spectrum analyzer becomes a fixed-tuned receiver. In this state, the bandwidth is equal to the resolution bandwidth. Signal amplitude variations are displayed as a function of time. To avoid loss of signal information, the resolution bandwidth must be as wide as the signal bandwidth. To avoid any smoothing, the video bandwidth must be set wider than the resolution bandwidth.

Index